David Brenner

**PRACTICAL PLUMBING
DESIGN GUIDE**

PRACTICAL PLUMBING DESIGN GUIDE

James C. Church, P.E.
*Former Vice President and Chief Sanitary Engineer
Syska & Hennessy, Inc., Consulting Engineers;
Past International President
American Society of Sanitary Engineering*

McGRAW-HILL BOOK COMPANY
New York St. Louis San Francisco Auckland Bogotá
Düsseldorf Johannesburg London Madrid Mexico
Montreal New Dehli Panama Paris São Paulo
Singapore Sydney Tokyo Toronto

Library of Congress Cataloging in Publication Data

Church, James C, date.
   Practical plumbing design guide.

   Includes index.
   1. Plumbing — Handbooks, manuals, etc. I. Title.
TH6125.C49   696'.1   78-18823
ISBN 0-07-010832-3

Copyright © 1979 by McGraw-Hill, Inc. All rights reserved.
Printed in the United States of America. No part of this
publication may be reproduced, stored in a retrieval system,
or transmitted, in any form or by any means, electronic,
mechanical, photocopying, recording, or otherwise, without
the prior written permission of the publisher.

1234567890   HDHD   7865432109

The editors for this book were Tyler G. Hicks and
Esther Gelatt, and the production supervisor was
Thomas G. Kowalczyk. It was set in typewritten form and printed
in 8½ x 11 inch size to facilitate its use in design offices
on drafting tables where typewritten specifications are a frequent
companion to the pencil, calculator, and T-square.
Printed and bound by Halliday Lithograph Corporation.

CONTENTS

PREFACE xi

CHAPTER 1. GENERAL DESIGN PROCEDURES..................... 1-1
    A. Information to be obtained and items to be checked at the start of a project............ 1-1
    B. Procedure for the design of plumbing work and the preparation of plumbing drawings........ 1-1
    C. Items to be checked when completing a project..................................... 1-4

CHAPTER 2. PLUMBING, FIRE PROTECTION, AND SITE SYSTEMS....................................... 2-1
    A. General... 2-1
        1. Conformity with requirements 2-1
        2. Principles of design 2-1
        3. Preliminary plumbing utility loads 2-5
        4. Energy conservation possibilities in plumbing systems 2-8

CHAPTER 3. SITE WORK...................................... 3-1
    A. General... 3-1
        1. Principles of design 3-1
    B. Drainage systems.................................... 3-6
        1. General 3-6
        2. Storm water sewer systems 3-13
        3. Sanitary sewer systems 3-15
    C. Water supply systems............................... 3-19
        1. General 3-19
        2. Domestic water supply systems 3-22
        3. Fire-protection systems 3-23
    D. Gas systems.. 3-26
        1. Principles of design 3-26
        2. Piping 3-27

CHAPTER 4. BUILDING WORK.................................. 4-1
    A. General... 4-1
        1. Principles of design 4-1
    B. Drainage systems................................... 4-9
        1. General 4-9
        2. Storm-water drainage systems 4-10
        3. Sanitary drainage and vent systems 4-18
        4. Laboratory waste-water drainage and vent systems 4-24
        5. Garage drainage and vent systems 4-27
        6. Radioactive waste-water drainage and vent systems 4-29
        7. Highly infectious waste-water drainage and vent systems 4-29
    C. Domestic water supply systems..................... 4-30
        1. General 4-30

     2. Street-pressure systems 4-48
     3. Boosted-pressure systems 4-48
     4. Hot-water systems 4-65
     5. Chilled-drinking-water systems 4-96
     6. Distilled-water systems 4-99
     7. Demineralized (Deionized) water systems 4-100
     8. Reverse osmosis (RO) water systems 4-102
     9. Water treatment 4-102
    10. Nonpotable-water systems 4-104
D. Irrigation Systems.................................. 4-104
     1. Principles of design 4-104
     2. Pipe sizing 4-108
     3. Piping 4-108
E. Swimming pools...................................... 4-109
     1. Principles of design 4-109
     2. Piping 4-112
     3. Equipment 4-113
F. Decorative pools and fountains.................... 4-117
     1. Principles of design 4-117
     2. Piping 4-128
     3. Equipment 4-128
G. Gas systems.. 4-132
     1. Principles of design 4-132
     2. Pipe sizing 4-135
     3. Piping 4-137
H. Fire standpipe systems: combined fire standpipe and sprinkler systems (New York City).......... 4-137
     1. Principles of design 4-137
     2. Piping 4-141
     3. Equipment 4-142
I. Sprinkler systems: combined sprinkler and fire standpipe systems................................. 4-145
     1. Principles of design 4-145
     2. Piping 4-148
     3. Equipment 4-149
J. Fire extinguishers................................. 4-151
     1. Principles of design 4-151
K. Carbon-dioxide fire-extinguishing systems........ 4-152
     1. Principles of design 4-152
L. Halon fire-extinguishing systems................. 4-154
     1. Principles of design 4-154
M. Dry-chemical fire-extinguishing systems.......... 4-155
     1. Principles of design 4-155
N. Foam fire-protection systems..................... 4-155
     1. Principles of design 4-155
O. Compressed-air systems............................ 4-156
     1. Principles of design 4-156
     2. General 4-160

      3. Pipe sizing 4-160
      4. Piping 4-162
      5. Equipment 4-162
P. Vacuum air systems................................. 4-167
      1. Principles of design 4-167
      2. Pipe sizing 4-170
      3. Piping 4-172
      4. Equipment 4-173
Q. Oral vacuum systems................................ 4-174
      1. Principles of design 4-174
      2. Pipe sizing 4-174
      3. Piping 4-174
      4. Equipment 4-174
R. Vacuum cleaning systems............................ 4-177
      1. Principles of design 4-177
      2. Pipe sizing 4-178
      3. Piping 4-179
      4. Equipment 4-181
S. Medical gas systems................................ 4-181
      1. General 4-181
      2. Oxygen systems 4-183
      3. Nitrous oxide systems 4-188
      4. Nitrogen systems 4-190
T. Central soap systems............................... 4-191
      1. Principles of design 4-191
      2. Piping 4-192
      3. Equipment 4-192
      4. Outlets 4-193
      5. Individual dispensers 4-193
U. Gasoline systems.................................. 4-193
      1. Principles of design 4-193
V. Insulation....................................... 4-193
      1. Principles of design 4-193
W. Vibration isolation............................... 4-195
      1. Principles of design 4-195
X. Plumbing fixtures................................. 4-198
      1. General 4-198
      2. Water closets 4-199
      3. Urinals 4-200
      4. Lavatories 4-201
      5. Sinks 4-203
      6. Flushing-rim-service (clinical) sinks 4-204
      7. Slop (service) sinks 4-204
      8. Mop sinks 4-204
      9. Laundry trays 4-205
      10. Drinking fountains (water coolers) 4-205
      11. Bathtubs 4-206
      12. Showers 4-207
      13. Emergency-use fixtures 4-208

    14. Bedpan washer: Sterilizers 4-208
    15. Can washers 4-208
    16. Hydrotherapeutic equipment 4-209
    17. Laboratory trim 4-210
    18. Fixture trim 4-211
    19. Toilet accessories 4-211

CHAPTER 5. CALCULATION FORMS............................. 5-1
    A. Electrical Connection Data Sheets................ 5-2
    B. Plumbing Connections Required for H.V.A.C. Equipment... 5-3
    C. Storm Water Drainage (Site) (3 Sheets).......... 5-4
    D. Storm Water Drainage Sizing (Leaders)........... 5-7
    E. Storm Water Drainage Sizing (House Drains)...... 5-8
    F. Drainage Sizing (Sanitary House Drains)......... 5-9
    G. Water Service Losses, Building Losses and Pump Head Calculations (4 Sheets)................ 5-10
    H. Water Riser Sizing............................... 5-14
    I. Water Main Sizing................................ 5-15
    J. Hot Water Recirculation System Sizing........... 5-16
    K. Gas Riser Sizing................................. 5-17
    L. Gas Main Sizing.................................. 5-18
    M. Laboratory Gas Sizing............................ 5-19
    N. Compressed Air Sizing............................ 5-20
    O. Vacuum Sizing.................................... 5-21
    P. Oxygen Sizing.................................... 5-22
    Q. Nitrous Oxide Sizing............................. 5-23

APPENDIX. UNITS OF MEASURE AND S.I. CONVERSION FACTORS   A-1

INDEX   I-1

LIST OF TABLES

Table	3-1	Pipe availability for sewers.................	3-9
Table	3-2	Site sewer pipe selection schedule..........	3-10
Table	3-3	Pipe material...............................	3-12
Table	4-1	Minimum wall requirements for plumbing fixtures.....................................	4-5
Table	4-2	Minimum wall requirements for plumbing fixtures (dry wall construction)............	4-7
Table	4-3	Lengths of expansion loops...................	4-8
Table	4-4	Lengths of typical branch takeoffs..........	4-8
Table	4-5	Load values assigned to fixtures............	4-43
Table	4-6	Estimating demand...........................	4-44
Table	4-7	Hospital water factors......................	4-45
Table	4-8	Office buildings, schools, and apartment water factors................................	4-45
Table	4-9	Water-pipe sizes (without flush valves).....	4-46
Table	4-10	Water-pipe sizes (flush valves).............	4-46
Table	4-11	Water-pipe sizes (1/2-gpm lavatory faucets).	4-47
Table	4-12	Factors for sizing house pumps...............	4-60
Table	4-13	% of volume withdrawn from hydropneumatic pressure tanks with varying pressure differentials................................	4-64
Table	4-14	Heat loss Btu/(h)(lin ft)...................	4-74
Table	4-15	Friction loss in ft/100 ft of hot water cir. piping..................................	4-75
Table	4-16	Maximum allowable lengths of dead-leg hot-water branches..........................	4-76
Table	4-17	Hot water demand in fixture units...........	4-77
Table	4-18	Hot-water demands and use for various types of buildings.................................	4-79
Table	4-19	Hot-water demand per fixtures for various types of buildings...........................	4-80
Table	4-20	Plumbing fixture hot-water ratings..........	4-81
Table	4-21	Proposed hospital plumbing fixture ratings..	4-89
Table	4-22	Residential hot-water sizing................	4-92
Table	4-23	Hot-water requirements for apartments.......	4-93
Table	4-24	Hot-water requirements for motels and hotels......................................	4-94
Table	4-25	Hot-water requirements for dormitories......	4-95
Table	4-26	Data for sizing gas piping..................	4-136
Table	4-27	Pressure loss (psi) per 100 ft in 50 compressed-air piping.......................	4-161
Table	4-28	Air-pressure loss, psi in 100 ft of clean commercial steel pipe.......................	4-165
Table	4-29	Pressure loss per 100 ft in vacuum air piping......................................	4-172
Table	4-30	Pressure loss in oxygen piping..............	4-187
Table	4-31	Pressure loss in nitrous oxide piping.......	4-189
Table	4-32	Pressure loss in nitrogen piping............	4-191

LIST OF FIGURES

Figure 3-1	Kutter's pipe flow chart	3-7
Figure 3-2	Manning's pipe flow chart	3-8
Figure 3-3	NBS nomograph 31	3-18
Figure 4-1	Hunter curves (up to 250 fixture units)	4-39
Figure 4-2	Hunter curves (up to 3,000 fixture units)	4-40
Figure 4-3	Expanded hunter curves (up to 30,000 fixture units)	4-41
Figure 4-4	Hazen and Williams pipe flow chart (friction table)	4-42
Figure 4-5	Steel house tank construction	4-62
Figure 4-6	Volumes in horizontal hydropneumatic pressure tanks	4-63
Figure 4-7	Modified Hunter curve for hot-water flow rate (up to 400 fixture units)	4-78
Figure 4-8	Modified Hunter curve for hot-water flow rate (up to 3,000 fixture units)	4-78
Figure 4-9	Angle spray design factors	4-127
Figure 4-10	Dew point conversion chart	4-166
Figure 4-11	Friction losses in oral vacuum piping	4-176
Figure 4-12	Friction losses in vacuum cleaning piping	4-180

PREFACE

This guide is not a beginner's theoretical text book but rather a book of useful design information gained from 40 years of practical experience designing plumbing and fire-protection systems for large and complicated buildings.

It is intended as a guide for anyone designing the many different plumbing and fire-protection systems for buildings and their sites. It contains design criteria that have been used successfully for many years in large and complicated buildings.

It can serve well as a design guide for either a large or small consulting engineering office, or as a valuable aid to an individual designer with an elementary knowledge of basic plumbing systems designing plumbing and fire-protection systems for all types of buildings.

It is basic and can be used in practically any locality, provided applicable local codes, local practices, and the requirements of the underwriters for the particular project are followed as overriding requirements, and provided a check is made of the proper domestic water-supply piping material for the particular water supply feeding the project. As local codes vary greatly, this design guide is a supplement to the local code.

It includes practically all the different systems that might be included in the plumbing and/or fire-protection contracts of a building. While probably no one building will contain all of these systems, buildings such as hospitals and research laboratories will contain many of them.

For the user's convenience, the guide has separate chapters covering project site work and work inside the building.

In addition to helping the designer, the information contained herein should help a contractor understand the workings of the systems (particularly the unusual or special systems) that he is installing. Similarly it should aid the inspector to understand these systems.

It can also serve as a reference and guide in teaching courses in advanced plumbing design.

It does not cover earthquake design requirements. For necessary provisions to withstand earthquake conditions in a particular locality, check the local codes and refer to the National Fire Protection Association Standard 13.

It also does not cover the specialized requirements of process piping systems.

The design recommendations contained herein should provide systems that will protect the public health and provide good sanitation, following the motto of the American Society of Sanitary Engineering: Prevention Rather Than Cure.

The guide was originally written as the Office Standard of Syska and Hennessy, Inc., Consulting Engineers of New York City. They have graciously allowed its publication in the interest of helping all the plumbing designers throughout the country.

CHAPTER 1. GENERAL DESIGN PROCEDURES

A. Information to be obtained and items to be checked at the start of a project.

   1. Location of the building or buildings on the site.
   2. Determine size, location, and depth of all adjacent available sewers, water, and gas mains. Investigate the requirements of local departments and/or utility companies, even the minor details, and confirm all information obtained. Failure to go into this thoroughly or to confirm information received can lead to embarrassing controversies when the project is under construction. If the locality is not too far distant, a telephone call to get or recheck information is worthwhile and should be made when necessary.
   3. Ascertain available water pressure at a given elevation.
   4. If the project is an alteration or an addition to an existing building, check existing services and equipment capacities.
   5. Determine the need for a fire standpipe system in the building.
   6. Check on the need for sprinklers in the building.
   7. Find out the need for special systems such as: gas, compressed air, vacuum air, distilled water, demineralized water, reverse osmosis water, medical gases, or other special systems.
   8. Check the local codes, if any.
   9. Verify with the electrical project engineer the current characteristics for motors, controls, and heaters, and the maximum size of motors that can be started across the line.
   10. Resolve with the HVAC project engineer the availability of steam or hot water for heating the domestic hot water.
   11. Obtain a detailed kitchen layout and equipment list.

B. Procedure for the design of plumbing work and the preparation of plumbing drawings.

   1. Count the plumbing fixtures and estimate the following:

      a. Sizes of house sewers.
      b. Size of water services.
      c. Size of gas service.
      d. Hot water load (tank size and necessary makeup).

   2. Determine if sump pumps or ejector pumps are required.

3. Determine if constant-pressure booster pumps or gravity or pneumatic tanks are required.
4. Note the presence or absence of sufficient slop sinks, fixtures for the handicapped, and drinking fountains or water coolers; and check any lack with the architect.
5. Ascertain whether drinking fountains will be supplied from a central or subcentral chilled water system or whether drinking fountains must be provided with local or integral cooling units.
6. Notify the electrical project engineer of the electrical requirements of the plumbing work.
7. Notify the HVAC project engineer of the domestic hot water heating requirements.
8. Ascertain from the HVAC project engineer his drain, water, and gas requirements. Great care should be exercised to work closely with the project engineer on all projects, so that:

    a. All water and gas requirements are satisfied by valved connections in the plumbing contract within 10 ft developed length of their equipment inlets, as the HVAC contract normally covers only the last 10 ft of piping to the equipment.
    b. Necessary floor drains, funnel drains, etc., are provided reasonably adjacent to all HVAC equipment requiring drains. All drainage piping between the equipment outlets and the floor drains, funnel drains, etc., will normally be provided in the HVAC contract.

9. Locate roof drains, leaders, main stacks, fire standpipe risers, sprinkler risers, etc.
10. Check with the HVAC and electrical project engineers regarding space conditions for risers, fire-hose cabinets, etc., and for required machinery room space.
11. Check the structural drawings for space conditions affecting the plumbing work.
12. Obtain minimum ceiling heights from the architectural drawings or architect and check to see that the plumbing work clears these heights.
13. Note lack of proper pipe spaces and wall thicknesses for the plumbing work and check this lack with the architect.
14. In general, piping should be run as direct as possible. However, except for piping buried underground, all piping should be run parallel to and at right angles to the walls, partitions, etc., and should be

neatly grouped in parallel lines.
15. In general, piping should be run to clear steel and concrete beams. Where absolutely necessary, piping may be run through beams. Where it is necessary to clip beam flanges or run piping through the web of steel beams or through concrete beams, permission from the structural engineer must be obtained and confirmed; and all such special conditions should be clearly noted on the drawings.
16. Note piping rising within a story as "rise." Note that rising to the story above as "UP." Piping dropping within a story should be noted as "drop." That dropping to the story below should be noted as "Dn." Piping at the ceiling should be noted as "at ceiling" when exposed and as "in ceiling" when concealed. Piping under the floor, other than obvious fixture drain lines, should be noted as "under floor," "at ceiling below," or "in ceiling below," as required.
17. Riser diagrams should be drawn at 1/4-in scale whenever possible (practical), but never less than 1/8-in scale. Water service, hot water heaters, sump pumps, ejector pumps, water pumps, tanks, etc., should be shown on riser diagrams.
18. All three phase motor controllers (starters), remote control stations, alarm panels, remote alarms, remote pressure switches, float switches, and electrode control units should be located and noted on the drawings.
19. Unless the architect or owner has a standard list of symbols that they request be used, plumbing drawings should use standard office symbols and abbreviations.
20. All information received from or given to the architect, owner, local authorities, etc., either in person or on the telephone, should be immediately confirmed by memorandum with copies sent to individuals involved.
21. When attending project meetings, detailed notes should be taken of all items of discussion pertaining either directly or indirectly to the plumbing work. After the meeting, a memorandum should be written confirming the details discussed with a copy sent to the others present at the meeting.
22. After a project has been scheduled and/or started, the project engineer should record by memorandum the origin of any change or addition to the work, the time involved (man days), and any resultant delay in completion, and have this memorandum approved before proceeding with the work involved.

23. Decisions affecting a trade other than one's own should be avoided when requested by the architect and/or owner. Your answer should be that the information will be obtained from the particular project engineer involved when you return to the office, or if necessary, by calling the project engineer involved from the architect's or owner's office. The same procedure should apply to telephone discussions.

C. Items to be checked when completing a project.

1. North point on all plans.
2. Column numbers, room names, and numbers on all plans.
3. Floor elevations indicated at least on first floor and basement or cellar floor plans.
4. General notes and reference to them on each drawing.
5. Symbol list.
6. Completeness of titles and title block, and presence of PE stamp on all drawings.
7. Reference notes between plans and risers, and between plans and plot (site) plan.
8. Indicated elevations of house sewers leaving the building.
9. Indications of sillcocks around the building on appropriate floor plan and plot plan.
10. Plans against risers for agreement of pipe sizing and location of vent and water branches (high or low).
11. Plumbing plans against those of the heating, ventilating, air-conditioning, and electrical trades for space clearances.
12. That the electrical project engineer has connected up all heaters, motors, controls, alarms, etc., and provided all required heat tracing (to prevent freezing). Check for agreement on location of motor controllers (starters), and horsepower or kilowatts required.
13. That the HVAC project engineer has connected up all hot water heaters, etc.
14. That all HVAC required water connections have been provided and are in the proper locations.
15. That all HVAC required drains have been provided and are in the proper locations.

CHAPTER 2. PLUMBING, FIRE PROTECTION, AND SITE SYSTEMS

A. General.

1. Conformity with requirements.

    a. All work shall be designed in accordance with the requirements of all applicable codes, the requirements of all local authorities having jurisdiction, the requirements of the owner's fire insurance underwriters, and in accordance with latest good engineering practice.
    b. For projects without plumbing code jurisdiction, contact your supervisor for the code to be followed.
    c. The following authorities should be checked for jurisdiction, availability of services, and requirements:

        (1) Building department.
        (2) Sewer department.
        (3) Water department or water company.
        (4) Gas department or gas company.
        (5) Fire department.
        (6) Public health and/or public safety departments.
        (7) Hospital department.

    d. All work shall be designed in accordance with this design manual, unless modified by your supervisor.

2. Principles of design.

    a. Your supervisor should always be consulted on the design of all unusual items, and any other items or systems which are not an everyday part of plumbing systems.
    b. See listing of preliminary plumbing utility loads, item 3.
    c. It is imperative to maintain consistency in design and detail from project to project on the same site or for the same owner, because inconsistencies without definite justification are impossible to explain to an owner.
    d. When handling buildings in new localities, investigate local customs and local requirements. Not only read the code but try to find out something about local practices. ASSE could be of help in this respect. Many plumbing inspectors are ASSE members.

e. The group designing the site work, i.e., site drainage, treatment plants, water supply plants, and other site utilities, and the sprinkler group should be given as much advance notice as possible regarding the starting and finishing dates and particularly changes in dates on the projects on which they are participating. It should be remembered that they are working simultaneously on several projects and must coordinate their program in order to accomplish maximum service.

f. For projects on which the owner's standard specifications are used (with or without modifications), make sure that all items normally covered in the specifications are covered for that particular project. If not, they must be added to the specifications or drawings so that there is a complete whole.

g. When an architect insists on changing the wording of the introductory paragraph under "Scope of Work," or any of the "Definitions," the matter should be checked with your supervisor, because failure to do so can lead to serious legal consequences with respect to completeness of the specifications.

h. In writing the list of acceptable manufacturers, care should be taken to select manufacturers not only for the quality of the project involved but also for the acceptability of their product with respect to the specifications. All divergencies from the list of acceptable manufacturers either in initial writing of the specifications or later in submission must be checked with your supervisor. (It is not necessary to list manufacturers for hangers and pipe and fittings where the pipe and fittings are specified by ASTM or ANSI Standards.)

i. Before commencing work on preliminaries, the project should be carefully reviewed with your supervisor to set up the design criteria and the work that should be indicated on the preliminary drawings, because such things vary widely with the type of building and the purpose of the preliminary.

   (1) General tendency has been to show more than is necessary on the preliminaries, which needlessly runs up the job cost.

   (2) The extent of the work and the detail indicated should also be coordinated with the other trades so that all are compatible.

j. The calculations shall be complete, including indications of all assumptions made and notations of the origin of outside figures appearing in them. In other words, they should be so complete and detailed that they stand on their own merits and can be interpreted by anyone without an explanation from the writer.

   (1) All calculations shall be dated.
   (2) Void calculations should be clearly marked "void" when a new calculation is made.

k. Special form calculation sheets should be used. See Chapter 5.

   (1) One-half to two-thirds of the data is printed on the calculation sheets, which will save time.
   (2) The calculation sheets proceed in a step-by-step process, thereby eliminating the possibility of leaving out a step because of some distraction.

l. In doing a preliminary, calculations should be prepared on the maximum size of the basic systems, and used for the sizing of the major equipment, even though the basis for such work may be very approximate or hypothetical. This is important for showing the relation between the sizes arrived at in the preliminary and the sizes developed in the final design.

m. On taking over a project that has been started by another, the project engineer should check very carefully through the previous records of the project in order to acquaint himself with what has been done and said about the project previously, to be sure that there is no gap in the work of the two project engineers, i.e., as a result of both assuming that the other took care of a certain item. This is particularly important on government jobs, where specific formal submissions of varying items must be made, and a source of embarrassment if an item is forgotten.

n. Drawing production should be carefully planned in advance to avoid costly overdrafting and to improve the clarity of the drawings. Noting should be kept to a necessary minimum, remembering that riser diagrams supplement floor plans; and if a combination of the two clearly presents the picture, much of the noting on the floor plans is unnecessary. When a pipe runs in the normal manner that such a pipe does in normal plumbing, it is not necessary to note this. However, when a pipe runs abnormally, such a fact must be noted.

o. Structural information should always be transmitted to the structural engineer.

  (1) All verbal information should be confirmed in writing.
  (2) Information given should include sketches for pits, steelcuts, weights of equipment and size of bases, slab openings, etc. (Freehand sketches are good enough).

p. As automatic 1/2-hp three-phase starters are disproportionately expensive compared with the cost of the equipment controlled, and in many cases it is impossible to get three-phase motors on some of this equipment, the electrical project engineer should allow plumbing systems to use single-phase for 1/2-hp motors.

q. The "Electrical Connection Data Sheet" establishes the general voltages available on the project. See Calculation Form A, Chapter 5.

  (1) These should be specified for heaters, solenoid valves, relays, etc.
  (2) However, motors should be specified in accordance with the following NEMA standard motor voltages:

Motor voltage rating, V	System voltage
115	110 to 120
200	208 to 220
230	240
460	480
575	600

r. When discussing any major items with the architect

or the contractor after a project is in construction, ascertain whether the same questions have been previously discussed with the field observer. This is important so that they can not keep discussing an item with different people in the office until they get a satisfactory answer. Field observers should inform you of all major discussions and interpretations given the architect or contractor. Similarly, project engineers should inform the field observers of any discussions or instructions given the architect or contractor.

   s. This design guide does not cover special earthquake design criteria. For such requirements, refer to special design criteria for the earthquake area involved.

3. Preliminary plumbing utility loads. When required at the conceptual stage of a project, preliminary load requirements may be calculated as follows.

   a. Domestic water supply and sanitary drainage. Multiply the population by the following gallons per day (gpd) per person, except as otherwise noted.

Building type	Total water	Hot water
Apartment (average occupancy 2) (per apartment)	150	75
Residence (average occupancy 4) (per residence)	250	125
Dormitories (residence halls)	65	20
Office building (8-h worker)	20	3
Industrial plant (8-h shift)	35	15
Schools without cafeteria or showers (per student)	15	2
Schools with cafeteria, no showers (per student)	20	4
Schools with cafeteria and showers (per student)	25	10

(Continued on next page)

(Continued)

Building type	Total water	Hot water
Hospitals (per bed)	300	125
Cafeteria and restaurants (per meal)	10	4
Places of assembly	3	1/2
Hotels (per room)	75	50
Nursing homes (per bed)	75	30
Airports (per daily passenger)	5	1
Commercial laundry* (per pound of wash)	3.5	2

*Hospital, 15 lb/day/bed; hotel with restaurant, 12 lb/day/bed; hotel without restaurant, 8 lb/day/bed; nursing home, 7 lb/day/bed; apartment or residence, 14 lb/week/living unit.

Temperature of use of hot water: 120°F; except cafeterias and restaurants-140°F, and commercial laundries-180°F.

To obtain the total daily requirements, add cooling tower and other HVAC water requirements to the above.

To obtain the average flow rate (gpm), divide the gallons per day by 60x, the number of hours of normal building operation (use).

To obtain the peak flow rate (gpm), multiply the average flow rate by 2 1/2.

b. Fire flow requirements (gpm).

Buildings	Sprinklers and hose	Hose only	Duration
Light hazard	500	500	30 min
Ordinary hazard:			
Group 1	700	500	60 min
Group 2	850	500	60 min
Group 3	1,250	500	60 min

Light hazard: Office buildings, hospitals, schools, institutional buildings, nursing homes, residential, places of assembly, small libraries, dining rooms.

Ordinary hazard group 1: Parking garages, light-industrial, laundries.

Ordinary hazard group 2: Large libraries, commercial, light industrial.

Ordinary hazard group 3: Exhibition halls, repair garages, warehouse.

Assume a fire in only one building. Assume the building with the greatest requirements.

Add to building requirements 1,000 gpm for outside hose streams.

c. Gas.

   (1) For cooking in apartments and residences, use the following table.

Living unit or apartment	Percent	Btu/h
1	100	65,000
5	45	146,250
25	25	406,250
50	20	650,000
100	15	975,000
200	13	1,690,000
300	10	1,950,000

To obtain CFH/h for natural gas, divide by 1,000 Btu/ft^3 and for liquified petroleum (LP) gas, divide by 2,500 Btu/ft^3.

Add for hot water heaters 50,000 Btu/h and for gas dryers 35,000 Btu/h.

(2) For restaurants and laboratories, proportion the requirements from the calculations of a similar project.

(3) Or for restaurants use the following:

Type of facility	Btu/customer served
Coffee shops (counter only)	28,500
Coffee shops (table service)	30,168
Cafeteria	32,667
Regular restaurant	33,528

4. Energy conservation possibilities in plumbing systems.

   a. Use of 1/2 gpm lavatory faucet (adaptors for

existing buildings).
b. Reducing hot water temperature at lavatory faucets from 105° to 95°.
c. Single lavatory faucet with 1/2 gpm flow of tempered water (95°).
d. Use of siphon jet type urinals instead of blowout type (may require privacy shields by the architect between urinals).
e. Use of flush valves on water closets giving 4-gal flush instead of 4 1/2-gal flush.
f. Shutting down hot water circulating pumps at night and over weekends.
g. Bypassing central chilled water unit when incoming water is below 50°.
h. Shutting down chilled water circulating pumps at night and over weekends.
i. Adjusting central chilled water unit to deliver 50° instead of 45° water.
j. Adjusting water coolers to deliver 55° water instead of 50° water.
k. Selecting pumps that have considerable running time, with efficiency the main and cost the secondary consideration.
l. Use of gravity house tank and house pumps instead of constant-pressure booster pumps.
m. Design of display fountains with smaller displays or designing with small displays for everyday use and bigger displays for use on only special occasions.
n. Omit winterizing of display fountain. Shut down and drain in winter.
o. Use of ejector pumps instead of pneumatic ejectors (higher maintenance may result).
p. Use of waste heat preheaters before hot water heaters whenever possible.
q. Use of recirculating water systems on rotary liquid ring type air compressors and vacuum pumps.
r. Use of water-saver tank type water closets (uses 30 percent less water).

CHAPTER 3. SITE WORK

A. General.

1. Principles of design.

    a. The site designer should work closely with the building project engineer in establishing the building loads.
    b. In sizing the various systems make ample provisions for anticipated future buildings, additions to buildings, and additional paved areas.
    c. Connections to public mains and sewers must be made in accordance with requirements of local authorities and codes.
    d. Insist that the architect obtain accurate survey information before attempting to finalize outside work, because it is impossible to design this work correctly without accurate information on existing conditions and final grades.
    e. In routing the building utilities, determine if there are any utility interferences between the building and street pipe that you intend to connect to. (In taking over a job from someone else, do not assume that this has already been done; make sure that your routing is clear.)
    f. Indicate all existing underground work that is known in the area. This includes piping, valves, manholes, pits, electrical wiring, telephone, etc., whether being connected to or not. This is necessary to avoid problems in the field with regard to the relative location of the pipes being connected to.
    g. The requirements of local departments and utility companies should be very carefully investigated and worked out with the respective parties and confirmed in writing.

        (1) It is very important that all their requirements, even minor details, be confirmed with them and not guessed at, in order to avoid unpleasant controversies when the project is under construction.
        (2) Use the utility data sheets on pages 3-3, 3-4, and 3-5.

    h. Coordinate the routing of all outside utilities before work is placed on final drawings. This includes sewers, water mains, gas mains, electric and telephone work, steam, high temperature and

chilled water, etc.
i. Draw profiles for all storm and sanitary sewers. This improves the layout and also facilitates coordination with other utility crossings.
j. Where existing utilities are located in an area which will be regraded, determine if the utilities will have enough cover, be exposed, or even be under the structure. (Manhole covers, drainage grates, and valve boxes must be adjusted to the new grades. This should be indicated on the drawings.)

Job:_____

Date:_____

Sewer Data

1. Size, location, and depth (including manhole locations) of all sewers fronting on or serving the site? Sanitary? Storm water? Combined?

2. If no sewers, how to dispose of sanitary sewage? Storm water drainage? If dry wells or sewage disposal system, topographical map and type of soil at varying depths?

3. If street sewer manholes are required, by whom? If by sewer department, at whose cost? If by owner, any special standard detail?

4. Any special standard detail required for house sewer manholes?

5. Are street sewer connections preferred at or between manholes? If between manholes, are there existing spurs? Are there any minimum or maximum limitations on size or depth?

6. Any other special requirements?

Job:_____

Date:_____

Water Data

1. Size, location, and depth of all water mains and fire hydrants fronting on the site or serving the site?

2. Water pressure at a given elevation?

3. Water meter: by water company? Purchased from water company or by owner? Type? Strainer? Water meter pit? If so, by whom? Standard pit details?

4. Fire hydrants: by water company or by owner? Type?

5. Amount of work done by water company? Amount of work done by water company and paid for by the contractor?

6. Minimum cover required on water lines?

7. Any other special requirements?

8. Water analysis.

9. Type of pipe, fittings and joints normally used? Required?

Job: _____

Date: _____

Gas Data

1. Size, location and depth of all gas mains fronting on the site or serving the site?

2. Btu content and pressure of gas available?

3. Amount of work done by the gas company? Amount of work done by the gas company and paid for by the contractor?

4. Equipment furnished by the gas company and to be installed by the contractor?

5. Approximate load: Btu/h

6. Size of service?

7. Size of meter?

8. Any special gas company requirements regarding location of meter or others?

9. Type of pipe, fittings, and valves normally used? Required? Coating? Cathodic protection?

B. Drainage systems.

   1. General.

      a. Principles of design.

         (1) At the start of a project, contact the sewer department of the community involved for the availability of storm, sanitary, and combined sewers in the area. (Obtain drawings showing locations, sizes, and depths of all available sewers, also any rules and requirements connected with sewer work.)
         (2) All drainage should be by gravity, wherever possible.
         (3) Do not design combined site sewers.
         (4) Design piping carrying building drainage in accordance with the requirements of applicable codes.
         (5) Design storm and sanitary sewers, not covered by code, in accordance with the requirements of all authorities having jurisdiction, or in the absence of same, as hereinafter indicated.
         (6) Provide manholes as follows:

            (a) For every change in direction.
            (b) For every change in pipe size.
            (c) For every change in pipe pitch (slope).
            (d) For every major connection.
            (e) At least every 300 ft.

         (7) Where sewers of different pipe size enter or leave a drainage structure, lower the crown of the larger pipe to match the crown of the smaller pipe. Where this is not possible, check with your supervisor.
         (8) Where pipes of equal size enter and leave a manhole, the outlet pipe should be lowered 0.1 ft.
         (9) Drop manholes should be provided when the difference in inlet and outlet elevations is at least the distance required for the required fittings, and is at least 3 1/2 ft.
         (10) Piping must be of sufficient strength for the depth of cover. (See rating sheets, Tables 3-1 to 3-3.)
         (11) See the following for pipe carrying capacities:

            (a) Kutter's pipe-flow chart, Figure 3-1.
            (b) Manning's pipe-flow chart, Figure 3-2.

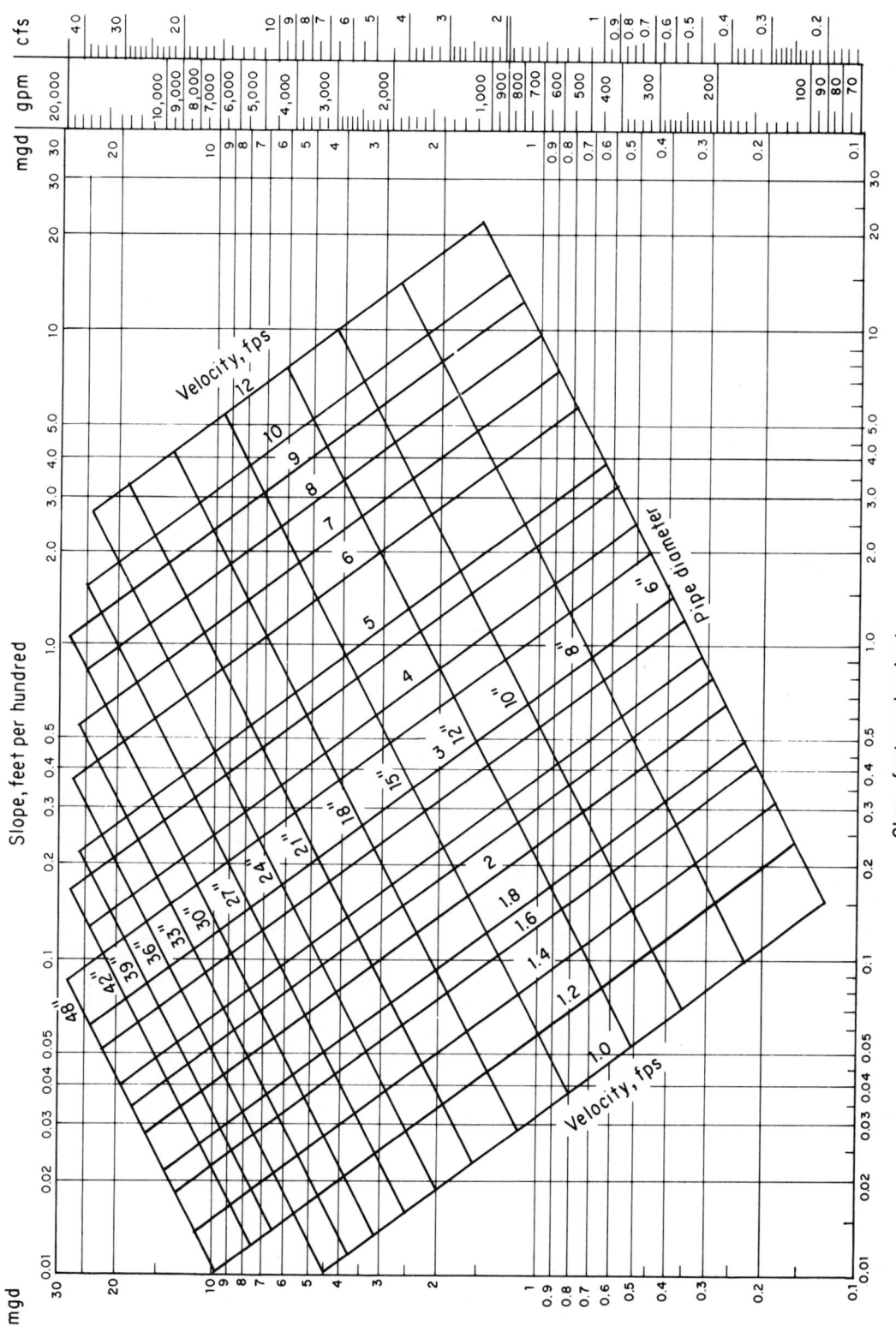

Fig. 3-1. Kutter's pipe flow chart. (N = 0.013)

From Technical Manual TM 5-814-1, Sanitary Engineering, Sanitary and Industrial Waste Sewers, August 1966, by the Department of the Army, USA.

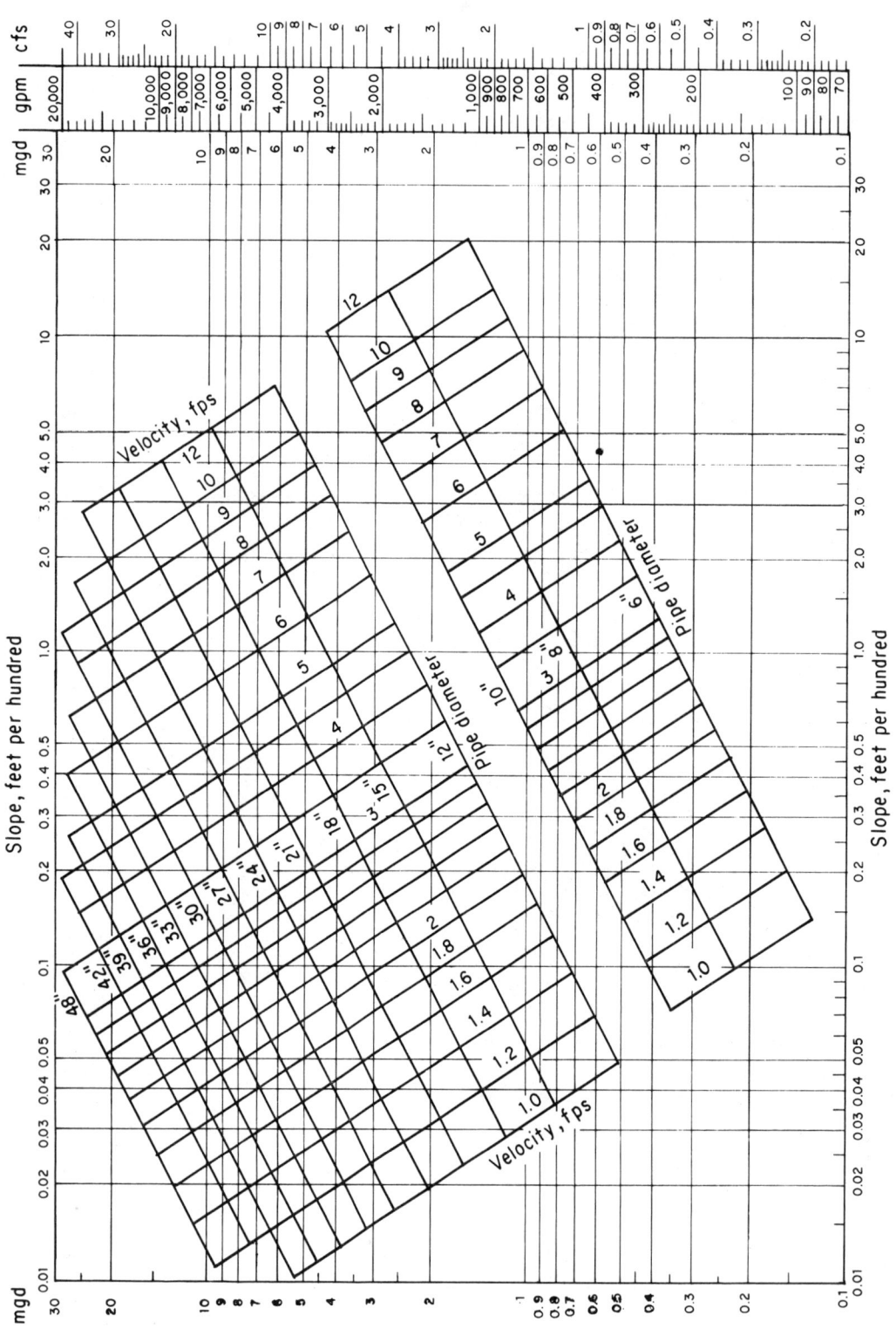

Fig. 3-2. Manning's pipe flow chart. (N = 0.013 for 12-in pipe and larger; N = 0.014 for 10-in pipe and smaller)

From Technical Manual TM 5-814-1, Sanitary Engineering, Sanitary and Industrial Waste Sewers, August 1966, by the Department of the Army, USA.

Table 3-1 Pipe availability for sewers

Material	1¼	1½	2	2½	3	4	5	6	8	10	12	14	15	16	18	20	21	22	24	27	30	33	36	42	48
Steel	X	X	X	X	X	X	X	X	X	X	X		X	X	X				X		X		X	X	X
CI* soil caulked		X			X	X	X	X	X	X			X												
DI* water					X	X		X	X	X	X	X		X	X	X			X		X		X	X	X
VCP*						X		X	X	X	X		X		X		X		X	X	X	X	X		
ACP*						X	X	X	X	X	X	X		X	X	X			X	X	X	X	X		
Plain concrete								X	X	X		X		X		X		X							
Reinforced concrete											X		X		X		X		X		X	X	X	X	X

*CI: cast iron.
 DI: ductile iron.
 VCP: vitrified clay pipe.
 ACP asbestos cement pipe.

Note to designer:

Steel: used only inside buildings, not below ground.
CI soil: can be used for all conditions. Lengths 5 or 10 ft.
DI water: can be used when CI is required over 15 in in size. Centrifugal cast 3 to 24 in 18 ft long.
VCP: lengths: 3 to 6 ft according to locality.
ACP:

Size, in	Lengths
4	5 ft class 1500; 5 ft and 10 ft class 2400 and 3300
5-6	5 ft and 10 ft class 1500, 2400 and 3300
8	10 ft class 1500; 13 ft class 2400 and 3300
10-36	13 ft all classes

Plain concrete: Lengths up to 8 ft.
Reinforced concrete: Lengths: standard
                      12 to 24 in; 4 or 6 ft
                      24 in; 4, 6, or 8 ft

Table 3-2 Site sewer pipe selection schedule*

Pipe size, in.	2 to 2.49	2.5 to 3	3+ to 4	4+ to 5	5+ to 6	6+ to 7	7+ to 8	8+ to 9	9+ to 10	10+ to 12	12+ to 14	14+ to 16	16+ to 18	18+ to 20
6	A / g													
8	B / E	a B / f / E	a b E / f	a b E / f	a b E / f	a b E / f	a B E / f	a B E / f / A	A B E / f	A B E / f g	A B E / g	A B E / g	A B F / g	A B F / g
10	A B / g / F	a B / f / E	a B E / f	a B E / f	a B E / f	a B E / f	a B E / f	a B E / f	A B E / g	A B E / g	A B F / g	A B F / g	A B F / g	A B F / g
12	AC / d BF	A / c B / E / g	a B / c E / f	a / bf E / c	a / b E / c / f	a B / c E / f	c A / B / E	c C / B / F	c A / B / F	d AC / BF / g	d AC / BF / g	d AC / BF / g	d AC / BF / g	d AC / BF / h
14	h G	g F	g E	g E	g E	g E	g E	g E	g F	g F	h F	h F	h F	h F
15	c B	c A / B	a / b / c	a / b / c	a / b / c	a / b / c	a B / c	a / c B	c A / B	c B / c	c B / c	d B / d	d B / d	d B / d C
16	h G	g F	g E	g E	g E	g E	g F	g F	g F	h F	h F	h F	j F	h F
18	c A / h / F	c A / h B / F	a bg F / c	a bg F / c	a bg F / c	a bg F / c	a bg F / c	a bg F / c	c A / h B / F	c A / h B / F	c A / c j B / F	c A / c j B / F	d AC / j BG / j	d AC / j BG / j
20	h F	h F	g F	g F	g F	g F	g F	g F	h F	h F	j F	j F	j G	j G

(Continued on next page)

Table 3-2 (Continued)

		Feet of cover										
Pipe size, in.	2 to 2.49	2.5 to 3	3+ to 4	4+ to 5	5+ to 6	6+ to 7	7+ to 8	8+ to 9	9+ to 10	10+ to 12	12+ to 14	14+ to 16	16+ to 18	18+ to 20
21	A B c	A B c	a b c	a b c	a b c	a b c	a b c	a b c	a b c	A B c	A B c	A B c	A B c	A B c d
24	A B G c j	A B F c h	a bg F b c	a bg F b c	a bg F b c	a bg F b c	a bh F b c	a bh F b c	a B F c h	A B F c j	A B G c j	A B G c j	A B G c	A B G c
27	A G c j	A G c j	a G c h	a G c h	a G c h	a G c h	a G c h	a G c h	a G c j	A G c j	A G c j	A G c	A G c	A G c
30	A G c	A H c j	a H c j	a H c j	a H c j	a H c j	a H c j	a H c j	a H c j	A H c	A H c	A H c	A H c	A H c
33	A H c	A H c	a H c j	a H c j	a H c j	a H c j	a H c j	a H c j	a H c	A H c	A H c	A H c	A H c	A J c
36	A H c	A H c	a H c	a H c	a H c	a H c	a H c	a H c	a H c	A H c	A J c	A J c	A J c	A J c
42	c	c	c	c	c	c	c	c	c	c	c	c	c	c
48	c	c	c	c	c	c	c	c	c	c	c	c	c	c

* For letter designations, see Table 3-3.

Table 3-3 Pipe Material

	Designation	
	Without cradle	With cradle
Vitrified clay pipe - extra strength (VCP)	a	A
Concrete pipe - extra strength (CP)	b	B
Reinforced concrete pipe - class IV (RCP)	c	C
Reinforced concrete pipe - class V (RCP)	d	D
Asbestos cement pipe - class 1500 (ACP)	e	E
Asbestos cement pipe - class 2400 (ACP)	f	F
Asbestos cement pipe - class 3300 (ACP)	g	G
Asbestos cement pipe - class 4000 (ACP)	h	H
Asbestos cement pipe - class 5000 (ACP)	j	J

Notes:

Dead loads: Marston's formula for loads on rigid conduits in trench conditions: $W_c = C_d W B_d^2$; W = 120 /CF; Ku = 0.165 max.

Live loads: Holl's integration of Boussineso's formula: $W_{sc} = C_s PF/L$; Impact factor F = 1.5, H-20 loading; L = 3.01; P = 20,000.

Trench: pipe bedding, class C, shaped bottom, load factor 1.5.

Trench: pipe bedding, class A, concrete cradle, load factor 2.8.

Safety factors: 1.0, reinforced concrete pipe; 1.5, vitrified clay pipe, concrete pipe, asbestos cement pipe.

Maximum allowable load on pipe: ASTM, three-edge bearing method to produce 0.01-in crack.

Trench width: OD, + 2 ft.

All pipe used between structures shall be the strongest pipe required by depth of cover as shown in Table 3-2 unless otherwise indicated.

2. Storm water sewer systems.

   a. Principles of design.

      (1) Obtain building requirements (both initial and ultimate) from the building project engineer. (Make ample provision in the piping for anticipated future building expansion.)
      (2) All building roofs, plazas, paved areas, and unpaved areas as required shall be drained by gravity to public sewers or other approved means of disposal. (Where gravity flow is impossible and pumping may be required, check with your supervisor.)
      (3) Where storm sewers are not available, the alternate means of disposal to be considered are:

         (a) Streams, lakes, or ponds.
         (b) Recharge basins.
         (c) Dry wells.
         (d) On grade.

      (4) There are computer programs available which can size the system.
      (5) Base storm-water drainage systems on at least a 10-year storm, for privately owned projects. For municipal work, verify this with the proper authorities.
      (6) Piping carrying solely site drainage can be sized using the following formula:

$$Q = ACI$$

Where Q = flow, ft^3/s

A = area, acres

I = rainfall rate (use 10-year storm), in/h

C = surface runoff factor

Roofs: $C = 1.00$

Pavement: $C = 0.90$

Grass: C = 0.30

(7) See Kutter's pipe-flow chart (Figure 3-1) or Mannings's pipe-flow chart (Figure 3-2) for pipe carrying capacity.
(8) Minimum size for site drains is 10 in, with size increased as required by the area drained.
(9) Select pipe size and pitch to provide a minimum velocity of 2 1/2 fps.
(10) Site drainage piping must have a minimum cover of 36 in.
(11) Site sewers serving site drains only may be looped through drainage inlets or catch basins (maximum of three) in lieu of manholes. Drainage from buildings should not be looped through drainage inlets or catch basins.
(12) The last drainage structure before connection to a combined sewer must be a hooded type.
(13) All curb-type drains must be hooded catch basins.
(14) Drains in unpaved areas may be catch basins or drainage inlets.
(15) Drains in paved areas must be catch basins.
(16) In working out trench drains with the architect, take care that trenches are wide enough to permit the installation of a proper type and size drain for its service. If the architect is unwilling to make the trenches the required widths, insist on the use of a cast-iron sectional trench with a drain in the bottom.
(17) Use round grates in grass areas. They are also suitable in paved areas, except where against a curb. They have the advantage of not being dropped into the catch basin or drainage inlet.
(18) Provide all low points on the site with a catch basin or a drainage inlet.
(19) Design piping so that the shortest routing is considered together with the depth of cut.
(20) Avoid placing manholes on very deep piping; try to locate the piping or manholes so that depths do not exceed 15 ft.
(21) Evaluate manhole sizes when multiple lines enter it.
(22) Consider direction of flow when connecting to manholes.

(23) Avoid placing pipes under known locations of trees. Review planting plan during design.
(24) Where possible, avoid placing piping under pavements.

b. Piping.

(1) Piping, unless otherwise noted, should be cast-iron soil pipe and fittings, or class 100 cast-iron or thickness class 2 ductile-iron bell and spigot water pipe and 250-psi fittings.
(2) Piping beyond 8 ft outside the building wall may be extra strength vitrified clay bell and spigot sewer pipe and fittings, extra strength or reinforced-concrete bell and spigot, and/or tongue and groove sewer pipe and fittings, or asbestos cement sewer pipe and fittings, as allowed by code.

(a) Minimum size of nonmetallic piping shall be 8 in.

(3) Use calculation forms C-1, C-2, and C-3, in Chapter 5, for sizing piping, when computer program is not used.

3. Sanitary sewer systems.

a. Principles of design.

(1) Obtain building requirements (both initial and ultimate) from the building project engineer. (Make ample provision in the piping for anticipated future building expansion.)
(2) All building house sewers should drain by gravity to public sewers or other approved means of disposal.
(3) Where sanitary or combined sewers are not available, alternate means of disposal to be considered are:

(a) Sewage treatment plant.
(b) Septic tank and tile field or leaching cesspools.

(4) Where other than public sewer disposal is required, consult your supervisor on the

requirements for this work.
(5) Piping should have a minimum cover of 36 in.
(6) Pipe size and pitch should be selected to provide a minimum velocity of 2 fps.
(7) Sewage pumping stations should be concrete structures (precast or poured in place), normally of the wet-pit-dry-pit type.
(8) Automatic air relief valves should be provided in trapped high points in the pump discharge piping and force mains.

b. Piping.

(1) Piping, unless otherwise noted, should be cast-iron soil pipe and fittings.
(2) Piping beyond 8 ft outside the building wall may be extra strength vitrified clay bell and spigot sewer pipe and fittings, or asbestos cement sewer pipe and fittings, if allowed by code. (Minimum size of nonmetallic piping shall be 8 in.)
(3) Sewage pump discharge piping inside the station should be galvanized standard-weight steel pipe with threaded cast-iron drainage fittings.
(4) Underground sewage pump discharge piping and force main should be coated cast-iron or ductile-iron bell and spigot water piping, or asbestos cement pressure pipe with cast-iron fittings, if its use is common practice in the area.

c. Equipment.

(1) Sewage ejector pumps.

(a) Pumps should be vertical wet-pit-dry-pit type or submersible type of the Flush-Kleen design.
(b) Pumps should normally be duplex (each full size). In large installations, and to provide for future loads, more than two pumps may sometimes be required.
(c) Calculate the pump head by adding the lift from low water in the pit to 1 ft above grade at the point of discharge (to ensure entry into a surcharged sewer) or the highest elevation of the discharge

line, whichever is greater, and the discharge pipe friction (including both individual pump discharge assembly and the common discharge piping).

(d) Low water in the pit is normally 18 in above the bottom for submerged pumps and dry-pit pumps, and the top of the motor for submersible pumps.

(e) Check the friction in the common discharge piping for both one- and two-pump operations. Normally friction for one-pump operation should be included in the specified pump head, provided there is not too great a drop in capacity with the friction of two pumps operating.

(f) The wet pit should have approximately 3 ft gross working depth (this will allow for controls and a reasonable operating cycle).

(g) Sewage ejectors should be sized to handle the peak anticipated flow as calculated from fixture units and National Bureau of Standards Nomograph 31, Curve 1, as shown in Figure 3-3.

(h) All controls should be the sealed electrode type, for each pump, or the "bubble" type.

(i) Compression tube high-water alarm actuating units should be provided.

(j) Where the lift is 35 to 40 ft or more, provide ejector pumps with external spring-type check valves instead of the regular swing checks used elsewhere.

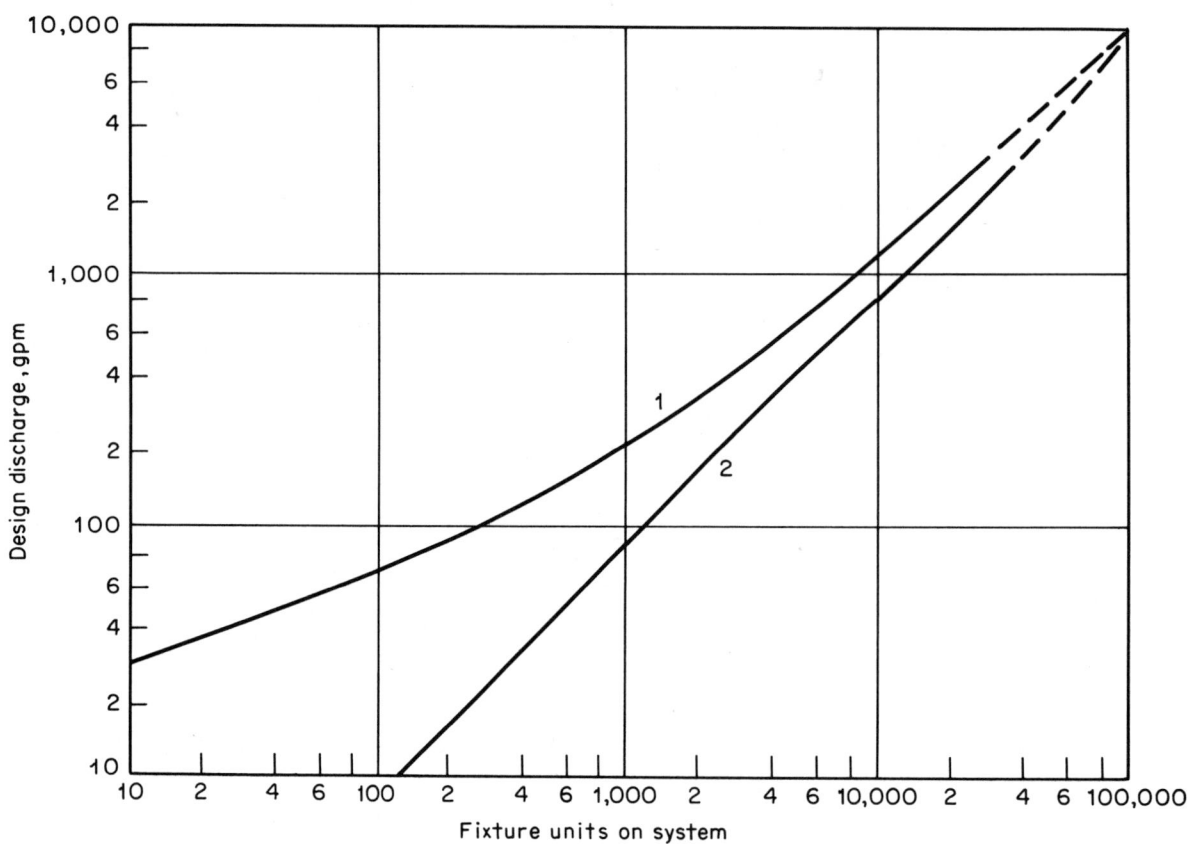

Fig. 3-3. NBS nomograph 31. Design-flow curves for plumbing-drainage systems. Curve 1 represents the peak discharge into the drainage system which will not be exceeded more than 1 percent of the time during periods of heaviest use. Curve 2, shown for purposes of comparison only, illustrates the average discharge into the drainage system during periods of heaviest use computed from the discharge characteristics of water closets.

Reprinted, by permission, from Air Conditioning, Heating & Ventilating (now Building Systems Design).

C. Water supply systems.

   1. General.

      a. Principles of design.

         (1) The local water department or water company should be contacted for the availability of a public water supply in the area.

            (a) Obtain drawings showing locations, sizes, and valving of water mains in the area and available pressure.
            (b) Obtain information about the water analysis (if necessary), the type and material of piping and joints, the type of meter, the location of the meter, and what work is normally done by the water department or water company.
            (c) In discussion with the local water department or water company, take care to ascertain the maximum possible street pressure as well as the minimum street pressure. They system design should be based on the minimum street pressure; and working pressures of equipment should be based on the maximum possible street pressure.
            (d) All projects involving street pressure sprinkler systems, street pressure fire standpipe systems, fire pumps, or domestic booster pumps require hydrant tests on the mains in streets that could be used to feed the site. Because these tests take time to get, they should be initiated as soon as possible at the start of a project. (Have a hydrant flow test made by the local water department or water company or the underwriters.)

(2) Where public water mains are available, extend service(s) into the site from same.
(3) Provide water services (both domestic and fire) to all buildings and to on-site fire hydrants as required.
(4) Sites having hospitals, laboratory-use buildings, or boiler plants should be provided with two services, feeding from separate mains wherever possible. If two separate mains are not available, on-site storage must be provided. Also, large projects such as office buildings, housing, and schools should be similarly provided with two sources of water.
(5) Where no public mains are available, water supply must be developed from wells, streams, or lakes with treatment plants as required and on-site storage of treated water. Consult your supervisor about requirements for this work.
(6) When a public water supply is not available, and wells are being considered, contact a local well driller for recommendations about possible capacities, quality, depths, types, and suggested locations.

   (a) There are basically two types of wells: one in unconsolidated (sand) formations having a double casing and a screen; and one in consolidated (rock) formations having a double casing and no screen. See AWWA Standards.
   (b) Well pumps are water-lubricated multistage turbine (actually multistage diffuser) pumps with electric-motor or -engine drive above, or with electric submersible drive.
   (c) Well pumps should be provided with electrode low water cutoff.
   (d) The pump head is calculated using the lift from the pumping water level, plus on allowance for future drop in level.
   (e) The water level in the well is usually much higher when the pump is not operating. If this high water level is below the top intermediate shaft bearing, provide a prelubricating device.

(7) Design potable water supply piping so that no used, unclean, polluted, or contaminated water can enter any portion of such piping from any tank, receptacle, equipment, or plumbing fixture by reason of back siphonage, suction, back pressure, or any other cause, either during normal use and operation thereof, or when any such tank, receptacle, equipment or plumbing fixture is flooded or subject to pressure in excess of the operating pressure in the water piping. Where this possibility exists, an air gap or an approved back-flow prevention device should be provided.

(8) Piping exposed to possible freezing should be insulated and provided with electric heating cable.

(9) Where site water mains are looped, sectionalizing valves should be provided.

(10) On-site fire hydrants should be located so that, together with street fire hydrants, all parts of all buildings will be covered by hose streams from two hydrants, using not over 500 ft of hose each. This coverage need be provided only for the outside of buildings provided with interior fire hose stations.

   (a) Locate fire hydrants around the buildings, preferably no closer than 20 ft nor more than 50 ft from the building.
   (b) Locate fire hydrants so that they are accessible to the fire department from an adjacent roadway. Review this with the fire department and architect.
   (c) The minimum size branch to fire hydrants should be 6 in and provided with a curb valve.

(11) Size piping on building loads (500 gpm minimum for fire standpipe system and the maximum estimated sprinkler flow) with an allowance of 1,000 gpm for fire hydrants.

(12) Ample provision must be made in the pipe sizing for anticipated future building loads.

(13) Services should be provided with approved type meters; and where there are two services to the site, a check valve must be provided in each service. (Compound-type meters should be provided with a straight pipe ahead of them

　　　　　　　equal to at least eight pipe diameters.)
　　　　(14) Water meters, back-flow preventers, etc.,
　　　　　　　should be located in pits or inside the build-
　　　　　　　ing as required by the local water department
　　　　　　　or water company. (Obtain all details and
　　　　　　　requirements for the meter installation from
　　　　　　　the local water department or water company.)
　　　　(15) Domestic and fire water services should nor-
　　　　　　　mally be split outside the building.
　　　　(16) Underground piping must have the minimum cover
　　　　　　　used in the area of the project. Requirements
　　　　　　　can be obtained from the local water depart-
　　　　　　　ment or water company.

2. Domestic water supply systems.

　　a. Principles of design.

　　　　(1) Building requirements (both initial and ulti-
　　　　　　　mate) should be obtained from the building
　　　　　　　project engineer.

　　　　　　　(a) Ample provision should be made in the
　　　　　　　　　　piping and in the equipment where practi-
　　　　　　　　　　cal, for anticipated future building ex-
　　　　　　　　　　pansion.
　　　　(2) Where no site fire mains with fire hydrants
　　　　　　　are being provided, the required site fire
　　　　　　　hydrants should be connected to the site
　　　　　　　domestic water main and normally have two
　　　　　　　2-1/2 in. connections and a pumper connection.

　　　　　　　(a) Refer to Section C.1.a.(10) for fire
　　　　　　　　　　hydrant requirements.
　　　　(3) Where the basic fire hydrants are connected
　　　　　　　to a site fire main, at least one fire hy-
　　　　　　　drant should be provided on the site domestic
　　　　　　　water main adjacent to a siamese on the site
　　　　　　　fire main to allow the fire department to
　　　　　　　boost the pressure in the site fire main.
　　　　(4) Where on-site storage must be provided, it
　　　　　　　should be in steel or concrete tanks, pre-
　　　　　　　ferably above ground. Do not use concrete
　　　　　　　tanks below ground without specific approval
　　　　　　　from the water department or water company,
　　　　　　　or where no public supply is involved, from
　　　　　　　the Department of Health

　　　　　　　(a) Refer to NFPA Standard No. 22, Water
　　　　　　　　　　Tanks, for construction of tanks.

(5) Booster pumping stations, when required, may be located in the building (designed under the building system) or in a separate building designed under the site work.

    (a) The site designer should work closely with the building project engineer as to the capacity and type of equipment involved.
    (b) Refer to Chapter 4, Section C.1.b. and C.3.b. Building Domestic Water Supply Systems for pipe sizing and further data on equipment.

(6) All building domestic water services and fire hydrant branches should be provided with curb valves.

(7) The maximum allowable velocity in the water piping is 8 fps, except the maximum velocity in pump suction connections, which is approximately 5 fps.

b. Piping.

(1) Unless otherwise noted, underground piping 3 in and larger should be cement-lined class 150 cast iron or thickness class 2 ductile iron bell and spigot water pipe and fittings.

(2) For underground piping beyond 8 ft outside the building walls, class 150 asbestos cement pressure piping may be considered if its use is common practice in the area.

(3) Underground piping 2 in and smaller should be type K copper tubing with cast brass flared type fittings.

(4) Underground valves shall be AWWA IBBM double-disc gate valves with curb boxes.

(5) Friction in cement-lined cast-iron, cement-lined ductile-iron, and asbestos cement piping should be calculated on the basis $C = 140$.

(6) For irrigation systems, refer to Section D in Chapter 4.

(7) For swimming pools, refer to Section E in Chapter 4.

(8) For decorative pools and fountains, refer to Section F in Chapter 4.

3. Fire-protection systems.

  a. Principles of design.

    (1) Obtain building requirements (both initial

and ultimate) from the building project engineer. Make ample provision in the piping for anticipated future building expansion.
(2) Design fire-protection systems in accordance with the requirements of the fire chief or the fire marshall, and the owner's insurance underwriters. (Systems should be tentatively designed on the basis of NFPA standards and good engineering judgement and submitted to the authorities for approval and/or comment.)
(3) For fire-protection standards, refer to NFPA Standards 13, Sprinklers; 14, Standpipe and Hose Systems; 20, Fire Pumps; 22, Water Tanks; and 24, Outside Protection.
(4) Materials and equipment must be approved by the underwriters.
(5) Where separate site fire mains are provided, they should be looped around the buildings with adequate sectionalizing valves.
(6) Provide separate site fire mains with a siamese adjacent to a fire hydrant connected to the site domestic water main or the street main to allow the fire department to boost the pressure in the site fire main.
(7) Provide fire hydrants on the site fire main. If the fire main is fed from a fire pump or siamese, the fire hydrants normally have only two 2-1/2-in connections. (Refer to Section C.1.a.(10), this chapter, for fire hydrant requirements.)
(8) For on-site storage, steel or concrete tanks, preferably above ground, must be provided. Elevated steel tanks on a tower may be required. (Refer to NFPA Standard 22, Water Tanks, for construction of tanks.)
(9) Fire pumping stations, when required, may be located in the building (designed under the building system) or in a separate building designed under the site work.

    (a) The site designer should work closely with the building project engineer concerining the capacity and type of equipment involved.
    (b) Refer to Sections H.3 and I.3, Chapter 4, for further data on equipment.
    (c) A jockey pump of generous capacity should always be provided.

(10) The system can have fixed connections from either the public (domestic) water supply or another water source such as a pond or lake, but not both.

(11) When the system is connected to the public water supply and a pond or lake exists on the site (which the fire department might use as suction for their pumpers pumping into a siamese), an approved back-flow preventer should be provided in the public water supply connection.

b. Piping.

(1) Unless otherwise noted, underground piping must be cement-lined class 150 cast-iron or thickness class 2 ductile-iron bell and spigot water pipe with 250-psi cast-iron fittings.

(2) For underground piping beyond 8 ft outside the building walls, class 150 asbestos cement pressure pipe with 250-psi cast-iron fittings may be considered, if its use is common practice in the area and it is approved by the underwriters.

(3) Underground cast-iron and asbestos cement piping subject to pressure from a fire pump or a siamese should be class 250.

(4) Inside piping should be standard-weight steel pipe with standard weight cast-iron threaded or flanged fittings, or Victaulic malleable iron fittings and couplings. Where fire pump discharge pressure may be over 175 psi, use extra heavy fittings. (Jockey pump piping and piping around a fire pump subject to water flow during periodic tests should be galvanized.)

(5) Underground valves should be post-indicator type or curb type, as required.

(6) For inside valves, refer to Sections H.2 and I.2, Chapter 4.

(7) Friction in steel-fire line piping can be calculated on the basis of $C = 120$.

(8) Friction in cement-lined cast-iron, cement-lined ductile-iron, and asbestos cement piping can be calculated on the basis $C = 140$.

D. Gas systems.

   1. Principles of design.

      a. Obtain building requirements (both initial and ultimate) from the building project engineer. (Make ample provision in the piping for anticipated future building expansion.)
      b. Gas is usually supplied from the utility company's street mains. When no street mains are available, supply must be from a liquified petroleum (LP) "bottled" installation.
      c. Contact the local utility company for the availability of street (natural) gas in the area. Obtain drawings showing locations and sizes of gas mains in the area. Obtain pressure, Btu/cu ft content of the available gas, and requirements regarding metering. (Gas services (natural gas) should be extended from the utility company's mains to all buildings requiring same.)
      d. When street gas is not available, LP bottled gas must be provided.

         (1) Contact the local LP gas supplier concerning the size of the tank recommended for the load involved and find out whether an underground or an aboveground installation is recommended and also who supplies the tank.
         (2) Review with the architect the tank location and determination of an underground or an aboveground installation.

      e. Design gas systems in accordance with the utility company's requirements, the requirements of applicable codes, and the requirements of NFPA Standards 54, National Fuel Gas Code, or 58, Liquefied Petroleum Gases, Storage and Handling.
      f. Gas service is usually low pressure (1/2 psi or less). However, it may sometimes be medium or high pressure and require reduction. Check with the utility company and the code about whether it is to be reduced inside or outside the building.
      g. Gas is usually metered inside the building, but it is sometimes metered on the site. Check with the utility company.
      h. Gas piping from the street main to the property line is normally provided by the utility company. Gas piping inside the property line is usually provided by the plumber. (Obtain from the utility

company the allowable pressure drop from the street main to the meter, from the street main to the regulator, from the regulator to the meter, whichever is applicable.)

2. Piping.

   a. Design piping in accordance with the requirements of the local code and the utility company (determine the extent of cathodic protection required).
   b. Piping must be made of materials recommended by the utility company.

      (1) Normally piping is black steel pipe with welded or "dreser" fittings.
      (2) Often it is coated "mill wrapped."
      (3) Often it is provided with cathodic protection.

   c. Piping from the street main to the meters is usually sized by the utility company. You are responsible for sizing the piping from the meter to the point of use.

      (1) When the meter is on the individual building service, the building project engineer will size the piping into the building.
      (2) When the meter is on the site service feeding a site distribution system, the site distribution system piping will be sized cooperatively by the building project engineer and the site engineer, as the total pressure drop allowed between the meter and the most distant point of use can be only 0.3 in of water.

   d. Curb valves should be lubricated plug valves with curb boxes.

CHAPTER 4. BUILDING WORK

A. General.

1. Principals of design.

    a. Make ample provision in the piping, and in the equipment, where practical, for anticipated future building expansion.
    b. Projects should be designed to achieve maximum water and energy conservation. Use of reduced-flow fixtures, minimizing temperatures, use of preheaters using waste heat or solar heat, time-clock control, selection of pumps with maximum efficiency, etc., should be considered.
    c. Insist that the architect obtain accurate survey information before attempting to finalize outside work, because it is impossible to design this work correctly without accurate information on existing conditions and final grades.
    d. In routing the building utilities, determine if there are any utility interferences between the building and street pipe that you intend to connect to. (In taking over a job from someone else, do not assume that this has been done by him; make sure yourself that your routing is clear.)
    e. All existing underground work that is known in the area should be indicated. This includes piping, valves, manholes, pits, electric and telephone work, etc., whether being connected to or not. This is necessary to avoid problems in the field of the relative location of the pipes being connected to.
    f. The requirements of local departments and utility companies should be very carefully investigated and worked out with respective parties and confirmed in writing.

        (1) It is very important that all requirements, even to minor details, be confirmed and not guessed at, in order to avoid unpleasant controversies when the project is under construction.
        (2) Use utility data sheets, Chapter 3.

    g. Carefully check each project for plumbing piping (both water and drainage) subject to freezing.

        (1) All drainage and water piping installed in exterior walls and other locations where freezing is a possibility must be provided

with heating cable and insulation.
(2) The economy of group wrapping with either electric heating cable or a heating tracer line should be considered with respect to groups of piping, as in a shaft on an outside wall.
(3) In unheated garages, all sanitary and storm water drainage piping, all water piping, and all wet fireline piping must be protected by heating cable (with an aluminum jacket over the insulation for at least 5 ft above the floor to protect from physical damage). However, systems of garage drainage need not be so protected, provided that no traps are installed on the floor drains.
(4) Review with your supervisor any questions about the need or possible need for freezing protection of any piping whatsoever on a project.
(5) Be sure to notify the electrical project engineer about the location and extent of all the electrically heat traced piping.

h. Water and drainage piping should not be run in elevator machine rooms; telephone rooms containing telephone equipment, relays, and terminal strips; electric rooms and closets containing exclusively equipment such as transformers, switchgear, motor control centers, panelboards or similar items of equipment; and in emergency generator rooms, except for piping directly associated with the generator unit.

(1) Elsewhere run no piping within 5 ft laterally of such electrical apparatus as motor control panels, switchboards, and electric motors, except for branch piping connecting to the pump.
(2) Where unavoidable, obtain approval from the electric department and arrange for a waterproof slab or provide a drip pan under the piping, carefully checking headroom requirements for equipment below. Indicate where drip pan is to spill.

i. Do not run water and gas piping under buildings because this creates conditions which may require difficult maintenance and repair and/or cause

4-2

hazardous conditions.

j. In locating vertical piping in pipe spaces and furred in around columns, be sure that there is sufficient space for the piping involved, considering the materials of the piping, interconnection of the piping, and the insulation on the piping. Even if shown on our drawings, if the piping cannot be fitted in the field, a good deal of time can be lost and construction cost can be incurred by field modifications of the design to make it workable.

k. Refer to Tables 4-1 and 4-2, which give minimum space requirements for piping behind fixtures (not including stack requirements) both for block walls and for stud walls with wallboard.

l. Design horizontal piping to be within the hung ceilings established by the architect, where they exist, and to maintain maximum headroom at all times. Carefully check for clearance above equipment and for any possible interference with the work of other trades.

m. Piping should maintain a minimum clearance of 1 in between fitting hubs (finished covering over fitting hubs on insulated piping), and between fitting hubs adjacent work.

n. Do not install piping in floor fill except where it cannot be avoided.

   (1) Protect water, gas, compressed air, and vacuum air piping in floor fill with #22 USSG galvanized sheet metal U-shaped covers.
   (2) Surround covers and drainage and vent piping in floor fill with 1/2 in of cement mortar.

o. Pack sleeves through floor slabs and fire walls with asbestos or rock wool. Through interior waterproofing, seal top with mastic.

p. Provide adequately for the normal expansion and contraction of the piping and for piping crossing building expansion joints.

   (1) On long main runs or across building expansion joints, indicate adequate elbow swings or expansion loops.
   (2) At connections of branches to mains, risers and equipment, indicate sufficient number of elbow swings. (Refer to Tables 4-3 and 4-4 for lengths of expansion loops and swings.)

q. In addition to the general statement in the specifications, indicate on the drawings where expansion joints or swings are known to be required. (Special expansion requirements may require special consideration, such as Vitaulic or Barco joints, lead, rubber, etc.)
r. Care should be given to indicate pipe-size changes on the contract drawings, also indicating where eccentric increasers or reducers are required. No shortcuts should be taken when showing pipe-size changes on main equipment hookups.
s. While the specifications state that our drawings are diagrammatic, it is imperative that what is indicated on the drawings be installed within the available space confinements of the structure and coordinated with the other trades so that there is no interference.
t. Items recessed in a wall, such as fire-hose cabinets, drinking fountains (water coolers), thermostatic valve cabinets, siamese, sill cocks, etc., and the piping associated therewith should be coordinated with the architect so that adequate wall thickness or double walls are provided.
u. For group valves and other controls concealed in hung ceilings, walls, fire spaces, and shafts (except in areas of liftout tile ceilings), provide access doors.

   (1) Locate access doors so that they will be as inconspicuous as possible. In fancy areas, check their locations and finishes with the architect.
   (2) Where the riser is in the pipe space behind a battery of fixtures, locate the access door in the wall immediately above the baseboard rather than in the ceiling to save on pipe drops.

v. Check with the architect about the desired finish on sill cocks, siamese, fresh-air inlets, etc.
w. Provide pressure and/or combination gauges on the discharge and suction of all pumps except hot-water and chilled-water circulating pumps, sump pumps, and sewage ejectors.
x. Provide thermometers on inlet and outlet of all hot-water heaters and preheaters, central chilled-water units, on hot-water and chilled-water return (circulation) piping, and outlet of master thermostatic tempering valves.

Table 4-1 Minimum wall requirements for plumbing fixtures, sheet 1

Type fixture	Block wall construction, combination and support	Rough thickness, single wall, in	Clear space double wall, in
Water closet:			
Floor type	Single	4	3
Floor type	Back to back and batteries	6-8	6-8
Wall hung	Single vertical fitting	10	8
Wall hung	Single back-to-back vertical fitting	14	12
Wall hung	Single and battery horizontal fitting		12
Wall hung	Single and battery back-to-back horizontal fitting		18
Urinal:			
Wall hung	Single	6	4
Wall hung	Single back-to-back, backing plates	8	6
Wall hung	Single back-to-back, chair carriers		12
Wall hung	Battery-backing plates	6	6
Wall hung	Battery-chair carriers		9
Wall hung	Battery back-to-back, backing plates		8
Wall hung	Battery back-to-back, chair carriers		12
Stall	Single	4	3
Stall	Battery and back-to-back	6	6
Pedestal	Single	4	3
Pedestal	Battery and back-to-back	6-8	6-8
Bathtub		4	3

Table 4-1 Minimum wall requirements for plumbing fixtures, sheet 2

Type fixture	Combination and support	Rough thickness, single wall, in[a]	Clear space double wall, in[a]
Lavatory	Single	4	3
	Single back-to-back	6	4
	Battery and back-to-back, backing plates and wall carriers	6	6
	Battery, chair carriers	8	6
	Back-to-back, chair carriers	8	6
Sink or laundry tub	Single	4	3
	Single back-to-back	6	6
	Battery and back-to-back, backing plates and wall carriers	8	6
	Battery and back-to-back, chair carriers	8	6
Slop sink		6	4
Flushing rim sink	Single	6	6
Shower	Single	4	3
	Battery stalls	6	4
	Gang	6-8	6
DF (proj) or water cooler		4	3
DF semi-recessed	Piping beside	b	9-11[b]
DF semi-recessed	Piping behind		12[c]
DF recessed	Piping beside		11[c]
DF recessed	Piping behind		14[c]

[a]No allowance is included for stack requirements.
[b]As above for 4 or 6 in partition.
[c]From face of finish wall to inside of back wall.

Table 4-2 Minimum wall requirements for plumbing fixtures (dry wall construction)

Type fixture	Combination and support	Space required, in[a]
Water closet:		
Floor type	Single and back-to-back	4
Floor type	Battery and back-to-back	12
Wall hung	Single and back-to-back residential carrier Smith 500	8
Wall hung	Single commercial carrier Smith 400	10
Wall hung	Single back-to-back commercial carrier Smith 400	13
Wall hung	Battery carrier Smith 200	16
Wall hung	Back-to-back carrier Smith 200	19
Lavatory	Single and back-to-back	6
	Battery and battery back-to-back	11
Urinal	Single	6
	Battery	12
Sinks	Single and back-to-back	6
	Battery	11
Flushing rim clinical sink	Single	10
Shower	Single	4
	Back-to-back	6
Slop sink		8
Drinking fountains (water coolers)		
Projecting	Stud support	4
Projecting	Chair carrier	6
Simulated semi-recessed with cooler	Stud support	4
Simulated semi-recessed with cooler	Chair carrier	6
Semirecessed	With and without cooler	6
Recessed	With and without cooler	12
Wal-Pak		12
Unit water cooler	Chair carrier	6

[a]Space required means space between the back sides of the two gyp boards. No allowance is included for stack requirements.

Table 4-3
The following chart should be used to establish the length of expansion loop:

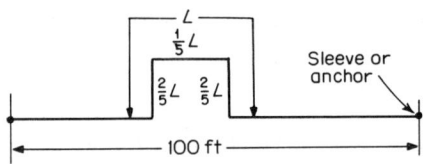

L feet per 100 feet

Pipe size	Building toilets, 80°ΔT, 120° system	Kitchens, 100°ΔT, 140° system	Laundries 140°ΔT, 180° system
1/2	6	7	8
3/4	7	8	9
1	8	9	10
1-1/4	9	10	11
1-1/2	10	11	12
2	11	13	14
2-1/2	12	14	15
3	14	15	16
4	15	18	19

Table 4-4 Typical branch takeoff

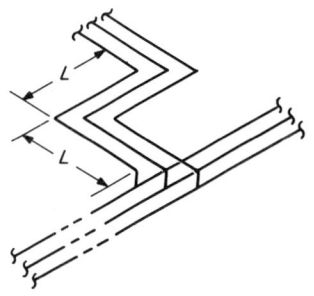

Pipe size	Minimum offset length L, ft
Up to 1-1/2 in	2
2 to 4 in	3
5 in and larger	4

4-8

B. Drainage systems.

1. General.

   a. Principles of design.

   (1) Contact the sewer department of the community involved for information about the availability of storm, sanitary, and combined sewers in the area. Obtain drawings showing locations, sizes, and depths of all available sewers, as well as any rules and requirements connected with sewer work.
   (2) Contact the local building department for specific requirements regarding house traps, rainfall rates, pipe materials and joints, and any other special local requirements.
   (3) Design drainage systems in accordance with the requirements of all applicable plumbing codes and good engineering practice.
   (4) Extend house drains out of the building as often as necessary to maintain required headroom in the building, consistant with the availability of site or street sewers on a particular side of the building.
   (5) No building should be designed with combined storm water and sanitary drainage systems. Each system must be completely independent out to the street or site sewer.
   (6) At least one of each duplex set of sump pumps and sewage ejectors should be fed from emergency electric-power circuits, if possible. Where constant drainage is critical and no emergency power is available, a third sump pump with a steam turbine or diesel engine drive should be provided.
   (7) Exercise great care to cooperate closely with the HVAC designer on all projects so that necessary floor drains, funnel drains, etc., are provided reasonably adjacent to all HVAC equipment requiring drains. Obtain from the HVAC designer a layout indicating the equipment and locations of floor drains. All indirect drainage piping between the HVAC equipment and these floor drains, funnel drains, etc., will be provided by the HVAC trade.

See Calculation Form B, Chapter 5.
(8) Do not run drainage piping at the ceilings of kitchens, food-preparation, food-serving, or food-storage area, if possible. (Check the code about this item.) If unavoidable and acceptable by code, drainage piping should be galvanized steel.
(9) Provide cleanouts as required by code and as required for proper maintenance of the systems. Indicate all required cleanouts. Extend all cleanouts in buried piping and in piping in hung ceilings up to cleanout deckplates in the floor above. Extend all cleanouts in piping concealed in walls out with wall plates. Where permitted by code, use 45° elbows (which do not require a cleanout) instead of one 90° elbow (which does require a cleanout).
(10) Where copper tubing or plastic pipe is used for drainage piping, provide expansion joints in stacks and leaders more often than provided in ferrous piping. In general, these expansion joints should be provided approximately every five stories for copper and as recommended by the manufacturer for the type of plastic used.
(11) Keep roof drains and stacks through the roof 12 to 18 in away from all parapet walls, building offsets, roof openings, etc., to allow for proper flashing. All stacks through the roof should be at least 4 in in size.
(12) Provide with electric heating cable and insulate any piping running through unheated areas or otherwise exposed to possible freezing. Where piping is located in outside walls, make provisions for getting some heat in the space, get assurance from the HVAC Department that space will not be below freezing or that piping must be heat-traced. Refer these conditions to your supervisor for review.
(13) If instructed by the owner or architect in writing to omit heat-tracing, use Victaulic couplings and fittings on storm water drainage piping in these locations, if allowed by the code or the plumbing inspector.

2. Storm-water drainage systems.

a. Principles of design.

   (1) All roofs, decks, balconies, canopies, terraces, etc., should drain by gravity to the sewer or other means of disposal. Plaza drains should have separate house sewers or a backwater valve before being connected to the building house sewer.
   (2) All paved areaways should drain by gravity to sump pits and be pumped into the main gravity system by duplex sump pumps.
   (3) All below-grade machinery-room floor drains and other clear water drips should drain by gravity to sump pits. Check the code for required separation of these sources into separate sumps and for connection to storm-water or sanitary systems.
   (4) Ample provision must be made in the piping for anticipated future building expansion.
   (5) The architect sometimes locates the roof drains. They should be checked to see if they are workable. This is particularly important with canopies, overhangs, and balconies. The ideal location is the quarter points. Keep the runs from the drains to the leaders as short as possible. In tall buildings, it pays to minimize the number of leaders with more horizontal piping on the upper floor ceilings.
   (6) Check for roof expansion joints because they create a bump in the roof, thus dividing the roof into fixed drainage areas.
   (7) Where the drains on lower roofs connect to the leaders serving higher roofs, drop them down beside the leader 2 or 3 ft and then connect them with an upright Y.
   (8) Where there are limitations on the flow allowed to public sewers and elsewhere (where local code permits), controlled-flow roof drainage should be considered as a matter of economy. The design should be in accordance with code requirements or the following, whichever is the stricter:

       (a) Use controlled flow only on nonpromenade roofs with parapets.
       (b) Obtain from a drain manufacturer curves for the community in which the building

is located. Do not use catalog tables. Use one weir (notch) per drain except for very large roof areas and heavy rainfalls.

(c) Use preferably a 25-year storm, and as a minimum, a 10-year storm.

(d) Allowable buildup:

Flat roof; 3-in water depth.

2-in pitch depth; 4-in water depth at drain.

4-in pitch depth; 5-in water depth at drain.

6-in pitch depth; 6-in water depth at drain.

(e) Maximum draindown time, 24 h.

(f) Maximum distance from the roof edge to the drains, 50 ft; and maximum distance between the drains 200 ft.

(g) Minimum two drains for a roof of 10,000 ft^2 or less; and at least three drains for larger roofs with limitation of approximate maximum of 10,000 ft^2 per weir notch.

(h) Arrange with the architect to place scuppers 4 in above the roof for flat roofs, 5 in for 2 in pitched roofs, and 6 in for 4- and 6-in pitched roofs above the elevation of the roof at the drain for pitched roofs. Provide at least one scupper for each 20,000 ft^2 of roof area.

(i) Inform the architect in writing to provide curbs at the doors and to extend the flashing above the overflow level.

(j) Inform the structural engineer about the loading on the roof due to the depth of water buildup.

(k) Use table on following page for piping sizing.

Size, in	gpm leaders	gpm horizontal drainage piping	
		1/8 in pitch	1/4 in pitch
3	67	34	48
4	143	78	110
5	261	139	196
6	423	222	314
8	911	478	677
10	1,652	860	1,214
12	.....	1,384	1,953
15	.....	2,473	3,491

For combination regular- and controlled-flow systems, convert the gpm to square feet and size by the normal storm water drainage tables.

(9) All drains on surfaces not intended for promenade should have dome strainers; others should have flat strainers with free area equal to twice the outlet area.
(10) Provide all areas exposed to rain, unless very small, with at least two drains.
(11) Do not use 2 in drains except for very small ledges, balconies, and canopies, or where 3 in cannot be accommodated in the construction.
(12) Plaza, terrace, and yard drains should normally be a minimum of 6 in to prevent clogging and preferably have locked hinged grates and buckets.
(13) In working out plaza, ramp, roadway and trench drains with the architect, insure that trenches are wide enough to permit installation of a proper type and size drain for its service. If the architect is unwilling to make the trenches the required widths, insist on the use of a sectional trench drain instead of a concrete trench with a drain in the bottom.

(14) Supply drains in areas subject to dirt and grit with a free-standing bucket in the drain body.
(15) Planting box drains.
    (a) Inside building: Dome-type roof drain with bronze mesh over the dome.
    (b) Outside (small areas): Perforated stand-pipe with a removable dome cap.
    (c) Outside (large areas): Free-standing bucket and low dome strainer, provided with an extension access to grade and a surface grate, and half tiles leading into the access extension.

(16) Base final drain selections on the physical structural conditions at the drain locations. Investigate the waterproofing conditions at all drains.
(17) Pipe pump gland leakage and bedplate drains to spill over adjacent floor drains.
(18) Foundation and/or subsoil drains, if required, should be drained by gravity to sump pits through sediment pits. Sediment pits should have a common wall with sump pits and be at least 3 ft wide to allow for cleaning.
(19) Foundation and/or subsoil drainage requirements and layout will usually be established by the structural and/or foundation engineers.

    (a) Determine at the beginning of the project if this is required.
    (b) Obtain expected rate of flow from the structural and/or foundation engineers for sizing of the sump pumps.

(20) Sump pits serving garage drains must be airtight and vented.
(21) Where discharge from garage drains runs into an adjacent stream or pond, provide oil separators. (Size separators for the normal flow (floor hosing down) with a full size over the top bypass to carry the fire flow.)
(22) Provide backwater valves in piping wherever a possibility of a backup exists.
(23) Provide interior drains and areaway drains with traps.
(24) Provide house traps only where required by code.

(25) See Calculation Forms D and E, Chapter 5, for sizing piping.

b. Piping.

   (1) Underground drainage piping should be cast-iron soil pipe and fittings up to 15 in. Over 15 in shall be class 100 cast-iron or thickness class 2 ductile-iron bell and spigot water pipe with 250-psi fittings.
   (2) Drainage piping inside the building should be cast-iron soil pipe, extra heavy, service weight, or No-Hub and fittings; or galvanized standard-weight steel pipe with threaded cast-iron drainage fittings. Where permitted by code, Victaulic fittings and couplings may be used, and they should be used if exposed to freezing. (Check code for allowable materials.)
   (3) Sump pump discharge piping, except underground, should be galvanized standard-weight steel pipe with threaded cast-iron drainage fittings. Where permitted by code, Victaulic fittings and couplings may be used.
   (4) Drainage piping from canopies, that is cast in columns, may be type L copper tubing or, if allowed by code, plastic.
   (5) Where allowed by code, plastic pipe and fittings may be used, if approved by your supervisor.

c. Equipment.

   (1) Sump pumps should normally be duplex (each full size) vertical submerged type in a 6 ft X 6 ft pit with approximately 3 ft gross working depth.
   (2) Self-priming type sump pumps should not be used unless requested by the owner.
   (3) Sump pumps should be increased in capacity above normal requirements to provide flooding protection of lowest areas of the building.

      (a) The extent of this protection must be evaluated in terms of the area involved, the nature of all occupancies of the areas, the potential possibility of a flood.

- (b) In small areas it may be desirable to have motors mounted on double-high motor-mount pedestals (30 in); and in very small subbasements of large buildings, it may be advisable to use submersible-type pumps.
- (c) Inlet piping and/or ventilated manhole cover (or grating) must be large enough to allow entrance into the pit of the peak capacity of the pumps.
- (d) All controls should be electrode type.

(4) Calculate the pump head by adding the lift from low water in the pit (normally 1 ft above the bottom for submerged pumps, and the top of the motor for submersible pumps) to 1 ft above grade outside the building or to the underside of the beams of the first floor above grade, whichever is greater, and the discharge piping friction (including both individual pump discharge assembly and the common discharge piping).

- (a) Increase the common discharge in size as necessary to allow for both pumps to operate simultaneously in emergencies.
- (b) Check the friction in the common discharge piping for both one- and two-pump operation. Friction for one-pump operation should normally be included in the specified pump head, provided there is not too great a drop in capacity with the friction of two pumps operating.

(5) Compression-tube high-water alarm-actuating units should be provided on sump pits.
(6) For submerged-type pumps, check the available headroom over the pit for adequate clearance for installation and removal of the pumps, where the pit is over 7 ft deep. If required headroom is unobtainable, use submersible pumps.
(7) Consider use of submersible pumps when pit depth is over 8 ft.
(8) Cellar-drainer-type sump pumps should be placed in 12 in X 15 in X 12 in depressions in the bottom of pits they are to drain. They must be specified to handle 200°F water, for condensate pits.

(9) Where the draining of other pits will considerably lower the drainage piping and increase the pit depth, consider providing a cellar drainer in the pit and pumping into the gravity piping.

(10) Remind the electrical project engineer to make all wiring to sump pumps of a waterproof type. Only electrode controls should be used where submersible sump pumps are used.

(11) Where the lift is 35 to 40 ft or more, provide sump pumps with spring-loaded check valves instead of the regular swing checks used elsewhere.

3. Sanitary drainage and vent systems.

   a. Principles of design.

   (1) All plumbing fixtures, drains, and equipment requiring drainage, above grade, should be drained by gravity to the sewer or other means of disposal.
   (2) All plumbing fixtures, drains, and equipment requiring drainage, below grade, should be drained by gravity to duplex ejectors and lifted into the gravity systems.
   (3) Make ample provision in the piping and ejectors for anticipated future building expansion.
   (4) In buildings over four stories, size house drains at the base of stacks generously to minimize the buildup of back pressure in the base of the stack.
   (5) In buildings over ten stories, connect plumbing fixtures on the grade floor to the ejector system rather than the gravity system.
   (6) In buildings over twenty stories, connect plumbing fixtures on the second and third floors above the house drain to substacks connecting independent of the main stack to the house drain.
   (7) In buildings with diverse occupancy in the top and the bottom portions of the building (usually requiring a radically different number of stacks and stack locations top and bottom), consider splitting the drainage system into two zones: one that serves the upper occupancy with a house drain at the ceiling below collecting the stacks and running down independently to the street sewer; another that serves the lower occupancy in a normal manner, with vents collected on the top ceiling of the zone and run through an offset rooof or the top roof. This will save on stack sizes and may eliminate need for substacks.
   (8) Where fixtures are scattered in a building, there is a tendency to make a "snake" of the stack to the point where the whole stack conception is lost. Stacks should run up through the building as directly as possible with minimal offsets only as required by architecture,

and stray fixtures should be connected to these stacks through branches.

(9) Ejectors.

   (a) Pneumatic sewage ejectors are the most maintenance-free; however, they require more horsepower than pumps, particularly in larger sizes and/or high heads, and are the most expensive.
   (b) Second choice is wet pit dry pit pumps.
   (c) The least expensive and most troublesome are submerged pumps.
   (d) All wet pit dry pit systems and submerged systems (unless handling only small-service employee toilet) should be the Flush-Kleen type.
   (e) Where energy conservation is a factor, pumps should be selected in preference to pneumatic ejectors, unless pneumatic ejectors are the owner's choice.

(10) Provide the building with ample stacks to serve the building requirements, maintain minimum offsetting of stacks, and avoid long branches.

(11) Provide required relief vents between soil and vent stacks every 10 stories, and for stack and house drain offsets and suds zones.

(12) Experience has shown that on tall buildings, the fresh-air inlet must be taken up several stories above the street level to avoid back-ups and foul odors discharging through the inlet. This height can vary from job to job. There is nothing in any code on the problem.

(13) Drainage piping serving gang showers shall be sized on 100 percent demand of the room rather than fixture units.

(14) Provide backwater valves in piping when there is a possibility of a backup, including branches to fixtures connected to pneumatic ejectors and located on the lowest level of the system close to the ejectors. Also provide backwater valves on all branches to floor drains, shower drains, and mop sinks connected to pneumatic ejectors and located on the lowest level of the system.

(15) In general, provide floor drains in the following locations:

  (a) Adjacent to all pumps, refrigeration compressors, air compressors, vacuum pumps, boilers, hot-water heaters, and air-conditioning equipment.
  (b) In kitchens near dishwashers, steam kettles, large refrigerators, and elsewhere as required.
  (c) In toilet rooms only on request of the owner or architect. Also check the code for requirements of floor drains in toilet rooms.

(16) Provide house traps and fresh-air inlets only when required by code.
(17) See Calculation Form F, Chapter 5, for sizing piping.

b. Piping.

(1) Underground drainage and vent piping should be cast-iron soil pipe and fittings.
(2) Unless otherwise noted, drainage and vent piping inside building should be cast-iron soil pipe (extra heavy, service weight, or No-Hub) and fittings, or galvanized standard-weight steel pipe with threaded cast-iron drainage fittings and threaded galvanized standard-weight malleable-iron vent fittings.
(3) Drainage piping at ceilings of kitchens, food-preparation, food-serving, or food-storage areas (if allowed by code); and drainage piping from kitchen facilities, unless underground, should be galvanized standard-weight steel pipe with threaded cast-iron drainage fittings.
(4) Ejector discharge, compressed air, and vent piping should be galvanized standard-weight malleable-iron vent fittings.
(5) Suction, surge, and house-tank overflow and drainage piping should be galvanized standard-weight steel pipe with threaded cast-iron drainage fittings.
(6) In all cases where the local code allows the use of 2-1/2-in size for drainage and vent piping above ground, this size should be indicated on the drawings wherever it is the proper size. This allows the contractor to use 2-1/2-in steel pipe, if he chooses, or

3-in cast-iron.
(7) Indirect wastes in kitchens, 1 in and smaller, should be type L copper tubing with solder joint cast brass or wrought copper fittings.
(8) Where allowed by code, plastic pipe and fittings may be used, if approved by your supervisor.

c. Equipment.

(1) Pneumatic sewage ejectors.

(a) Calculate the head by adding the lift from the bottom of the pit to 1 ft above grade outside the building or to the underside of the beams of the first floor above grade, whichever is greater and the discharge piping friction.
(b) Pot discharge is twice-rated gpm capacity.
(c) Interlock pots to prevent simultaneous discharge.
(d) Normally pots should be provided with individual air compressors; however, in large sizes, a stored-air system may be more economical in first cost.
(e) The table gives the minimum pneumatic ejector pit sizes:

Capacity, gpm	Size[a]	Approximate grout, in
30	7'0" X 9'0" X 3' 5"	1 3/4
50	7'6" X 9'6" X 3' 6"	1 1/2
75	8'0" X 10'0" X 3'11"	1 1/2
100	8'0" X 10'0" X 3'11"	1 1/2
150	8'0" X 11'0" X 4' 9"	1 1/2
200	9'6" X 12'0" X 5' 0"	1 1/2
250	10'3" X 12'6" X 4' 6"	2
300	10'3" X 13'0" X 5' 0"	2
400	13'0" X 14'0" X 6' 1"	2

[a]The depth indicated is the distance from the centerline of the inlet header to the bottom of the pit including an allowance for grout. The total required pit depth is this distance plus the distance

below the floor of the centerline of the inlet piping at the pit.
(f) Provide pits with a curb and railing around the top and 12" X 15" X 12" depression in one corner for use of a portable pump if ever needed.
(g) Pots should be cast iron.

(2) Sewage ejector pumps.

    (a) Calculate the pump head by adding the lift from low water in the pit to 1 ft above grade outside the building or to the underside of the beams of the first floor slab above grade, whichever is greater, and the discharge piping friction (including both individual pump discharge assembly and the common discharge piping).
    (b) Low water in the pit is normally 18 in above the bottom for submerged pumps and dry pit pumps and above the top of the motor for submersible pumps.
    (c) Check the friction in the common discharge piping for both one- and two-pump operation. Normally friction for one-pump operation is included in the specified pump head, provided there is not too great a drop in capacity with the friction of two pumps operating.
    (d) Pumps should normally be duplex (each full size) in a 6 ft X 6 ft pit with approximately 3 ft gross working depth.
    (e) Provide a round cast-iron basin in or instead of a concrete pit when required by code.
    (f) Self-priming-type sewage ejector pumps must not be used unless requested by the owner.
    (g) Pump pits must be sealed and vented through the roof.
    (h) Ejectors receiving drainage from other than plumbing fixtures should be sump pumps.

(3) Size sewage ejectors to handle the peak anticipated flow as calculated from fixture units and the curve published in the National

Bureau of Standards Nomograph 31, Curve 1, see Figure 3-3.
(4) Provide compression-tube high-water alarm-actuating units on ejector systems.
(5) All controls should be the sealed electrode type for each pump.
(6) For submerged-type pumps, check the available headroom over the pit for adequate clearance for installation and removal of the pumps, where the pit is over 7 ft deep. If required headroom is unobtainable, use submersible pumps.
(7) Consider use of submersible pumps when pit depth is over 8 ft.
(8) Where the lift is 35 to 40 ft or more, provide ejector pumps with external spring-type check valves instead of the regular swing checks used elsewhere.
(9) Grease traps.

   (a) Provide grease traps where required by code.
   (b) Provide grease traps for the waste from grease-producing kitchen equipment such as prewash sink, prewash section of dishwasher (not wash or rinse section unless required by code), pot washer, pot sinks, and floor drains serving kettles; and for grease-producing industrial or process equipment.
   (c) Locate grease traps as close to the equipment served as possible, but situated such as to provide adequate access for cleaning.
   (d) Where practical, several fixtures or pieces of equipment may be connected to one floor-recessed grease trap.
   (e) On projects where the HVAC equipment requires a Rotoclone, provide them with a water connection and floor drain. The recirculating water system (which is part of the HVAC system) will be in the HVAC contract to centralize the responsibility. As this recirculating system includes a grease trap (normally a plumbing item), give them the specifications for same to include in their specifications and give them any and all assistance necessary in

sizing and selecting this item.
- (f) Grease traps should be coated cast iron, sized in accordance with the manufacturer's recommendations, and provided with a flow-control device.
- (g) When a grease trap is recessed in the floor slab, provide an extension to finish flush with the floor. In pan construction, check the trap dimensions for fitting in the pan or arrange for flat slab construction. Provide for flashing where required.

4. Laboratory waste-water drainage and vent systems.

   a. Principles of design.

   - (1) Neutralize wastes from all fixtures and equipment where acids are or may be used before discharge to the sanitary drainage system.
   - (2) Neutralize by chemical reaction. Wastes should normally discharge to sumps filled with limestone chips or, for very large laboratories and industrial installations, to a treatment tank using injected sodium hydroxide (NaOH) solution as the neutralizing agent.
   - (3) Trap every fixture individually.
   - (4) Clear wastes (i.e., emergency-station drains) in the area of the laboratory waste system and remote from a sanitary system may discharge to the laboratory waste system.
   - (5) Size the system to include future anticipated expansion and to provide for flexibility of the laboratory areas.
   - (6) Provide individual acid neutralizing sumps for isolated sinks. Individual neutralizing sumps replace the fixture trap.
   - (7) Provide central acid neutralizing sumps for areas of numerous sinks; and locate in an accessible place for easy maintenance.
   - (8) All waste piping from laboratory sinks and cup drains (where acids might be used) to acid-neutralizing sumps and all vent piping for these fixtures should be of acid-resistant materials.

   b. Piping.

(1) Acid-resistant piping should be extra-heavy silicon iron (Duriron) pipe and fittings (either mechanical joint or hub and spigot, except only hub and spigot underground), or regular schedule borosilicate glass pipe and fittings with mechanical joints (extra heavy underground with protector jacket supplied by the manufacturer).

(2) Check with your supervisor before considering use of plastic piping.

(3) Fire-resistive polypropylene pipe and fittings may also be used where allowed by code.

(4) Glass piping in locations where it is subject to accidental damage, should be extra heavy and provided with protective guards, or a more sturdy pipe material substituted in that location.

(5) Glass piping through sleeves or construction must be completely protected by aluminum-foil tape.

(6) Heat-traced glass piping must be completely protected by aluminum-foil tape.

(7) Stacks and branch vents should be offset before going through the roof with acid-resistant plastic-lined threaded cast-iron or malleable-iron elbow at the base of the vent terminal.

(8) Vent terminals through the roof should not be glass pipe.

(9) See Calculation Form F, Chapter 5 for sizing of piping.

c. Equipment.

(1) Acid-neutralizing sumps should be sized on the following basis:

Diameter	X	Height		Users or sinks
12"		12"		1 sink
18"		12"		2 sinks
18"		24"	–	3-6
24"		36"	–	7-20
30"		57"	–	21-50
36"		70"	–	51-100
42"		52"	–	51-100
36"		6'0 "	–	40
42"		7'0 "	–	41-75
43"		5'9 "	–	41-90
48"		7'3 "	–	91-100
54"		8'3 "	–	101-150
60"		8'3 "	–	151-200
66"		8'3 "	–	201-250
72"		8'3 "	–	251-300
82"		8'6 "	–	301-400
90"		8'9 "	–	401-500
96"		8'11"	–	501-600
102"		9'0 "	–	601-700
108"		9'1 "	–	701-800
114"		9'3 "	–	801-900
114"		9'11"	–	901-1,000

(a) Sizes above the line are chemical stoneware. Sizes below the line are acid brick-lined steel.
(b) Users: the number of people likely to use the sinks and/or cup drains during a peak hour period.
(c) For classrooms: use the number of students.
(d) For research laboratories and hospital clinical laboratories: use the number of sinks and cup drains. One sink equals one cup drain.
(e) Underground acid-neutralizing sumps may be acid brick-lined concrete tanks of equivalent volume.
(f) In lieu of the chemical stoneware sumps listed above, approved plastic sumps of equivalent capacity may be used when allowed by code.

5. Garage drainage and vent systems.

   a. Principles of design.

   (1) Provide garage areas with floor drains to carry off the water dripping from the cars, to allow for hosing of the floor, and to remove the sprinkler discharge.
   (2) Connect garage floor drains to an independent system connecting to an oil separator and/or sediment trap before connection to the sump or main drainage system, as required by code.
   (3) Check code for venting requirements for garage drains.
   (4) Do not connect garage drains to a separate storm sewer, except where required by code.
   (5) Provide garage floor drains with a removable sediment bucket and trap, except where code or plumbing inspector allows omission of trap, as in unheated or outdoor garages.
   (6) Provide ramps from the outside for entering and leaving the garage with full-width trench drains at the base and partway up the ramp. Trench drains can be sectional cast-iron or domed drain in concrete trench provided by the G.C., depending on the architect's wishes.
   (7) In unheated garage areas, provide floor drains

with buckets having frost-protection holes in the bottom.

   (a) Use running traps instead of P traps, when underground.
   (b) Heat-trace and insulate P traps.

(8) Where a flow-rated oil separator (manufactured) is used, and an emergency flow of water far greater than the separator rating is possible, size piping for the emergency flow with a full-sized overflow bypass around the separator.
(9) Oil separators and sediment traps should be local standard where required by code.
(10) Floor drains should be located in conjunction with architect's floor pitches to give adequate drainage for hosing of the floor.
(11) Garage floor drains should be a minimum of 4 in outlet size.
(12) Horizontal piping serving two or more drains at the lowest level should be at least 5 in size for two drains and 6 in for three or more.

b. Piping.

(1) Piping should be the same as for storm-water drainage systems.
(2) Piping in unheated garages should be galvanized steel pipe with Victaulic fittings and couplings.

c. Equipment.

(1) Where manufactured oil separators are used, they should be coated steel and provided with accompanying waste oil tanks. Size separator for the estimated normal flow in gpm.
(2) Where oil separators discharge into a sump pit, the sump pit must be sealed and separately vented through the roof due to the possible presence of oil fumes.
(3) Where separators and waste oil tanks are buried, all maintenance openings (cover, cleanouts, tank suction connection, etc.) must be extended to the finished floor or grade as required to make them accessible.

6. Radioactive waste-water drainage and vent systems.

   a. Principles of design.

   (1) Waste-water drainage from fixtures or equipment which might be radioactive should be run separately from other drainage systems.
   (2) Treatment and/or disposition of radioactive waste-water must be determined on an individual basis based on the degree of radiation and the quantities involved, in accordance with the requirements of all authorities having jurisdiciton.
   (3) Conferences should be held with the owner and the local authorities regarding the design of this system and the disposal of its effluent.
   (4) Do not run horizontal drainage piping carrying radioactive waste-water on or in ceilings of occupied rooms, kitchens, food-preparation, food-serving, or food-storage areas unless adequate lead shielding is provided in the ceiling construction.

   b. Piping.

   (1) Piping should be the same as for sanitary drainage and vent systems.

7. Highly infectious waste-water drainage and vent systems.

   a. Principles of design.

   (1) Run waste-water drainage and vent piping from fixtures or equipment which might contain highly infectious waste material separately from other drainage systems; and effectively decontaminate it before disposal into the sanitary drainage system.
   (2) Determine treatment and/or disposal of highly infectious waste on an individual basis based on the degree of contamination and the quantities involved, after consultation with the owner on the degree of contamination.
   (3) Generally, high-temperature heating followed by cooling is required for the waste, and electric incineration of the vapors in the

                    vent terminal is required for the vents.
                (4) Depending on the degree of contamination,
                    wastes can be decontaminated by sanitizing
                    (heating to 200°F) or by sterilizing (heating
                    to 280°F).

        b. Piping.

            (1) Use stainless-steel drainage and vent piping
                for the untreated wastes.

C. Domestic water-supply systems.

    1. General.

        a. Principles of design.

            (1) Provide all fixtures and equipment requiring
                water with water in adequate quantities and
                at required pressures.
            (2) Where public (site) water mains are available,
                run service(s) into the building (site) from
                same. Connections to public water mains
                should be made in accordance with requirements
                of local authorities.

                (a) At the start of a job, ascertain from the
                    local water department or water company
                    the maximum possible street pressure as
                    well as the minimum street pressure.
                (b) The system design should be based on the
                    minimum street pressure, while working
                    pressures of equipment should be based on
                    the maximum possible street pressure.
                (c) Have a hydrant flow test made by the local
                    water department or water company or the
                    underwriters to verify actual pressures.

            (3) Provide all hospitals, laboratory-use build-
                ings, and boiler plants with two services,
                feeding from separate mains wherever possible.
                If two separate mains are not available, on-
                site storage must be provided. Also, large
                projects such as office buildings, housing,
                and schools should be similarily provided
                with two sources of water when possible.
            (4) Where no public (site) mains are available,
                water supply must be developed from wells,

streams, or lakes with treatment plants as required, and there must be on-site storage of treated water. Consult your supervisor about requirements for this work.

(5) Provide services with approved-type meters; and where there are two services per building, or buildings are cross-connected, provide a check valve in each service.

    (a) Provide compound-type meters with a straight pipe ahead of them equal to at least eight pipe diameters.
    (b) Check with the water company or water department about the type of meter to be used.

(6) Size each service for the full demand of the building.

(7) No potable water-supply piping should be designed so that used, unclean, polluted, or contaminated water can enter any portion of such piping from any tank, receptacle, equipment, or plumbing fixture by reason of back siphonage, suction, back pressure, or any other cause, either during normal use and operation thereof, or when any such tank, receptacle, equipment, or plumbing fixture is flooded, or subject to pressure in excess of the operating pressure in the water piping.

    (a) No plumbing fixture, device, or equipment should be connected to any domestic water supply when such connection may provide a possibility of polluting such water supply or may provide a cross connection between a distributing system of water for drinking and domestic purposes and water which may become contaminated by such plumbing fixture or device, unless an air gap or an approved backflow-prevention device is provided.
    (b) Check the code carefully for requirements, including check valves in some conditions.

(8) Size piping and equipment to include future anticipated loads and to provide for normal flexibility in laboratory areas.

(9) Valve all risers and branches as well as all equipment connections. Provide all risers with drain valves. Provide mains with sectionalizing valves at strategic locations. In hospitals, where practical, loop mains and generously provide them with sectionalizing valves. Valve all connections for future extension.

(10) Provide piping running through unheated areas or otherwise exposed to possible freezing with electric heating cable and insulation. Where piping is located in outside walls, there must be provisions for some heat in the space, assurance from the HVAC department that space will not be below freezing, or piping must be heat-traced.

(11) Required fixture and equipment pressures should be carefully investigated. Wall-hung water closets, showers, and dishwashers generally require 20 psi immediately at the fixture. Some fixtures and equipment such as low-tank silent water closets, some sterilizers, etc., may require more.

(12) Design systems so that a maximum pressure of 75 to 80 psi static (no flow pressure) is supplied to any fixture. Check code and owner's criteria for lower pressure requirements. A combination of master pressure-reducing valves can be used to accomplish this.

  (a) Where the inlet pressure at a master pressure-reducing valve station is under 100 psi, provide a valved bypass.
  (b) Where the inlet pressure at a master pressure-reducing valve station is over 100 psi, provide duplex valves. Size duplex valves so that each carries one-half the flow.
  (c) Master pressure reducing valves 2-1/2 in and larger cannot operate singly because they cannot handle very low flows; they must therefore be paired with a small-size similar valve to handle the low flows.
  (d) To avoid the use of four valves, use three valves instead of two.
  (e) Where the pressure reduction (inlet to

outlet pressure) exceeds a ratio of 3 1/2 to 1, two pressure reducing valves in series should be provided in order to avoid the cavitation likely to occur in greater single reductions.

(f) On cold-water down-feed risers, a branch pressure-reducing valve on the riser can serve the floors below. On cold-water up-feed risers and on all hot-water risers, branch pressure-reducing valves can be provided to feed two or three floors when the hot-water riser is in the pipe space behind the fixtures.

(g) There is a pressure differential between static (no flow) and flow conditions on the downstream side of pressure reducing valves. For master pressure reducing valves, assume this differential to be 5 psi. For branch pressure reducing valves, assume this differential to be 10 psi. Downstream pressures indicated on the drawings or specified should be flowing pressures.

(h) Strainers should be provided ahead of all pressure reducing valves.

(13) The maximum allowable velocity in the water piping should be 8 fps, with the following exceptions:

(a) Where hard water has been fully softened, maximum velocities in copper tubing and brass pipe should be 5 fps.

(b) The maximum velocity in pump suction connections (except hot-water and chilled-water circulators) should be approximately 5 fps.

(14) All cold-, chilled-, tempered-, and hot-water piping should be insulated.

(15) 12-in high air chambers, the full size of the branch, should be provided on the branch to each fixture. On branches controlled by solenoid valves, provide mechanical shock absorbers.

(16) Provide automatic air-relief valves at all high points in the circulated hot-water system that might become air-bound. These are

not required where the main feeds fixtures on the floor above. Where the need for these air-relief valves is known, they should be indicated on the drawings, leaving the generality of the specifications to cover only the unforeseen need for them.

(17) Make ample provision for expansion and contraction of the piping.

  (a) When laying out water-distribution systems, consider pipe expansion and contraction.
  (b) The piping that is affected most is the hot-water supply and return piping. It is subject to the greatest expansion and contraction, since the temperature differential between the installation at ambient (could be at freezing) temperatures and the operating temperatures is the greatest.
  (c) Since the system layout in the horizontal usually takes many turns and offsets and is not usually anchored to the structure intentionally, the piping system usually flexes and adjusts to the temperature changes without any known undue stress.
  (d) Occasionally the system layout is such that long straight runs are made which penetrate the slab at either end. If this occurs, either modify the run so that an offset can be incorporated or add an expansion loop.
  (e) The size of the expansion loop depends on the following factors:

    (1) Pipe material.
    (2) Length between confinements.
    (3) Possible temperature differential.

  (f) Refer to Table 4-3 for required expansion loop lengths.
  (g) Provide swing connections at connections to mains and risers. Refer to Table 4-4 for required lengths.
  (h) Anchor copper hot-water and hot-water return risers to the floor slab every six floors.

(18) All float valves must be of the fail-safe

type. Except for small valves (which must be the internal pilot type), all float valves should be the external pilot controlled type. Consider speed controls on these valves when tank turbulence or water hammer in the piping is a factor. The valves require at least 5 psi in the inlet piping to operate.

    (a) When makeup water is introduced through an air gap, provide an open- and closed-type (modulating type is not necessary) float valve with speed controls and with a throttling valve in the line ahead of it.

(19) Due to the high pressures usually present at the inlet of the float valve feeding an intermediate tank, this valve should be a comtination float and back-pressure type.

(20) Provide all float valves and float switches with enclosing pipes (stilling wells) to protect the float from turbulence and wave action.

(21) It is a good policy to provide Y strainers ahead of main automatic valves.

(22) Provide sillcocks around the exposed perimeters of the building and for promenade decks, spaced to give adequate coverage using not over 100 ft of hose. A sillcock should also be provided on the roof adjacent to cooling towers. Where a boosted pressure is required for cooling tower makeup, this sillcock should be connected to the same system.

(23) Provide hose bibbs in all machinery rooms, kitchens, animal rooms, and rooms requiring washing down, and in rooms with floor drains, where required by code.

    (a) Hose bibbs should be combination hot and cold, where requested by the owner and/or where required by the room use.

    (b) Where hose bibbs and/or hot and cold hose bibbs are required in public spaces, they should be in locked wall boxes, similar to sillcocks without the antifreeze feature.

    (c) All hose bibbs must be provided with vacuum breakers.

(24) Some can washers require at least 40-psi water pressure for proper operation. The availability of this required pressure should be checked for the location of the can washer; and where this is not available, an attempt should be made to relocate the can washer to an area of higher pressure, if at all possible. If this is not possible, a manual booster pump should be provided to give the required increase in pressure.

   (a) For the Dean single can washer, provide 1-in cold- and hot-water branches.
   (b) For Dean double can washer provide 1-1/4= in cold- and hot-water branches.
   (c) For the Bestov or Air Void can washer, provide 1-in branches.

(25) Give special attention to the water supply for special-events buildings, where athletic events with intermissions are held, because the demand is at least twice the Hunter-curve reading during these intermissions. Where a boosted pressure system is required, it should be a constant-pressure system with a small pump or possibly a hydropneumatic pressure tank for everyday use. For this condition, check with your supervisor.

(26) On projects in New York City, using street steam, it is required by law that the steam condensate be used for cooling-tower makeup and only the remaining requirements by city water. This requires the HVAC to have an open collection tank. On projects where they have an open collection tank, provide the cooling-tower makeup there instead of at the cooling tower, thus saving on size of the house tank and the house pumps.

(27) See Calculation Form G (4 sheets), Chapter 5, for calculating service losses, remaining pressures, and pump heads.

b. Pipe sizing.

   (1) Peak flows in the piping should be calculated and the pipe sized by:

      (a) Using fixture units (Table 4-5) and the

Hunter curves (published in National Bureau of Standards Report B145-79) Figures 4-1, 4-2, and 4-3; or Table 4-6; modified by the factors listed in Tables 4-7 and 4-8; and proper friction tables for the pipe material used.
    (b) Computer program for sizing domestic water piping.

  (2) In providing water connections for the HVAC contractor or others, the size of the connection should be at least as large as the connection on the equipment; or larger if required to deliver the gpm required with reasonable pressure loss. If the size requested seems to be abnormally large, the matter should be discussed with the HVAC project engineer.
  (3) See Calculation Forms H and I, Chapter 5, for sizing piping.

c. Piping.

  (1) Except where you have done previous projects using the same water supply, consult a chemical consultant about recommendations for piping materials. This information is often obtained for both plumbing and HVAC systems.
  (2) Unless otherwise noted, underground piping 2 in and smaller should be type K soft copper tubing.
  (3) Unless otherwise noted, underground piping 3 in and larger should be cast-iron or ductile-iron bell and spigot water piping.
  (4) For underground piping beyond 8 ft outside the buildings walls, asbestos cement (Transite) pressure piping may be considered, if its use is common practice in the area.
  (5) Generally inside the building, cold-water piping 5 in and larger should be galvanized standard-weight steel pipe with threaded galvanized standard-weight malleable-iron or flanged galvanized standard-weight cast-iron fittings, or Victaulic malleable-iron fittings and couplings.
  (6) Generally inside the building, cold-water piping 4 in and smaller and all hot-water piping should be type L copper tubing with (95-5) solder joint cast brass or wrought

copper fittings, or standard-weight red brass pipe with threaded standard-weight cast brass fittings. In New York City, copper tubing must be brazed (silver soldered) and type TP copper tubing can be added as another option.

(7) When the water is very hard, all piping inside the building should be galvanized steel.

(8) Where allowed by code, plastic pipe and fittings may be used, if approved by your supervisor. Where piping is insulated, fire and smoke ratings of the insulation supercede those of the pipe and fittings.

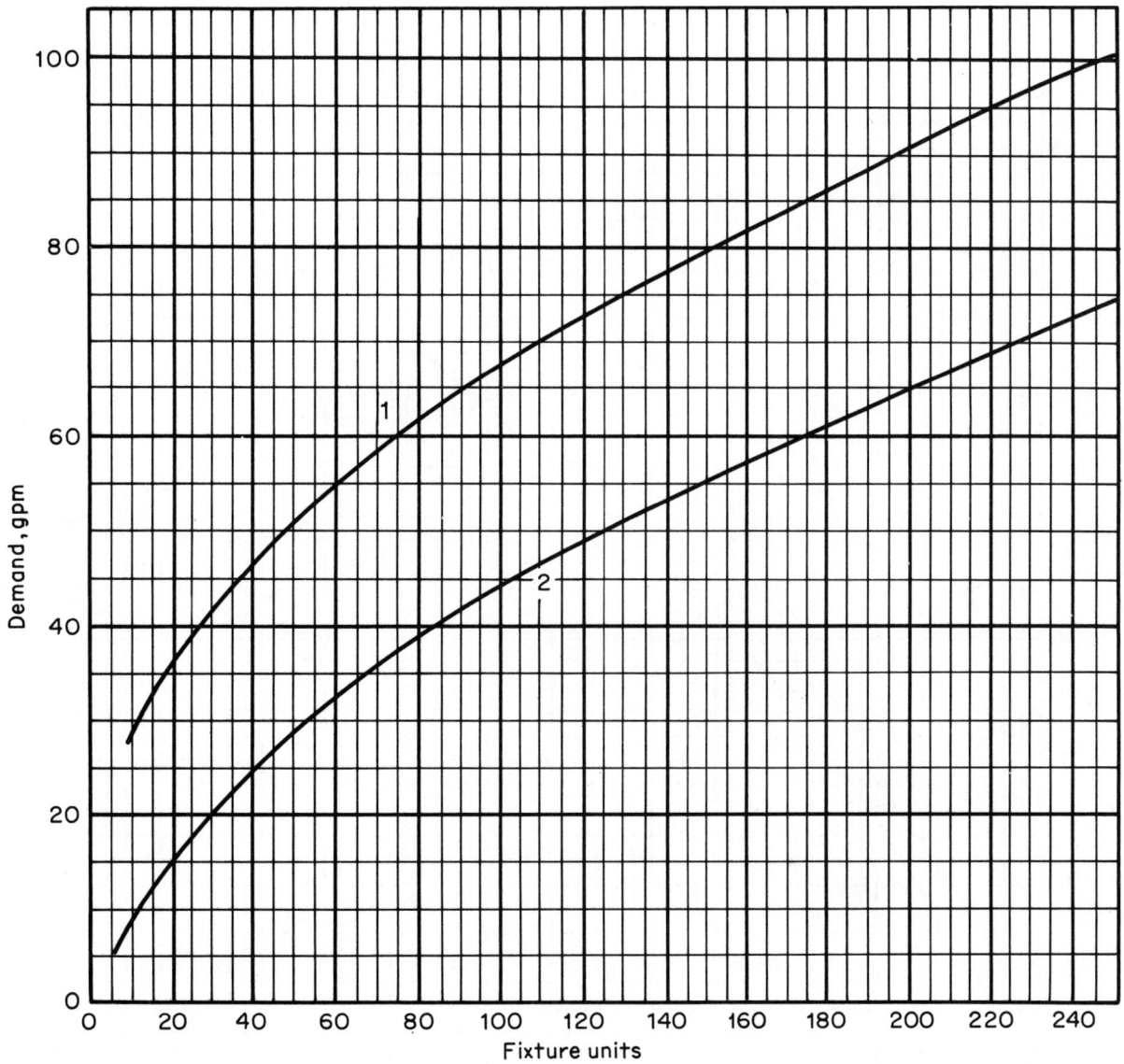

Curve 1 for system with flush valves
Curve 2 for system with flush tanks

Fig. 4-1. Hunter curves (up to 250 fixture units).

From Engineering Manual, Part V, Chapter 4, March 1946 by the Department of the Army, USA.

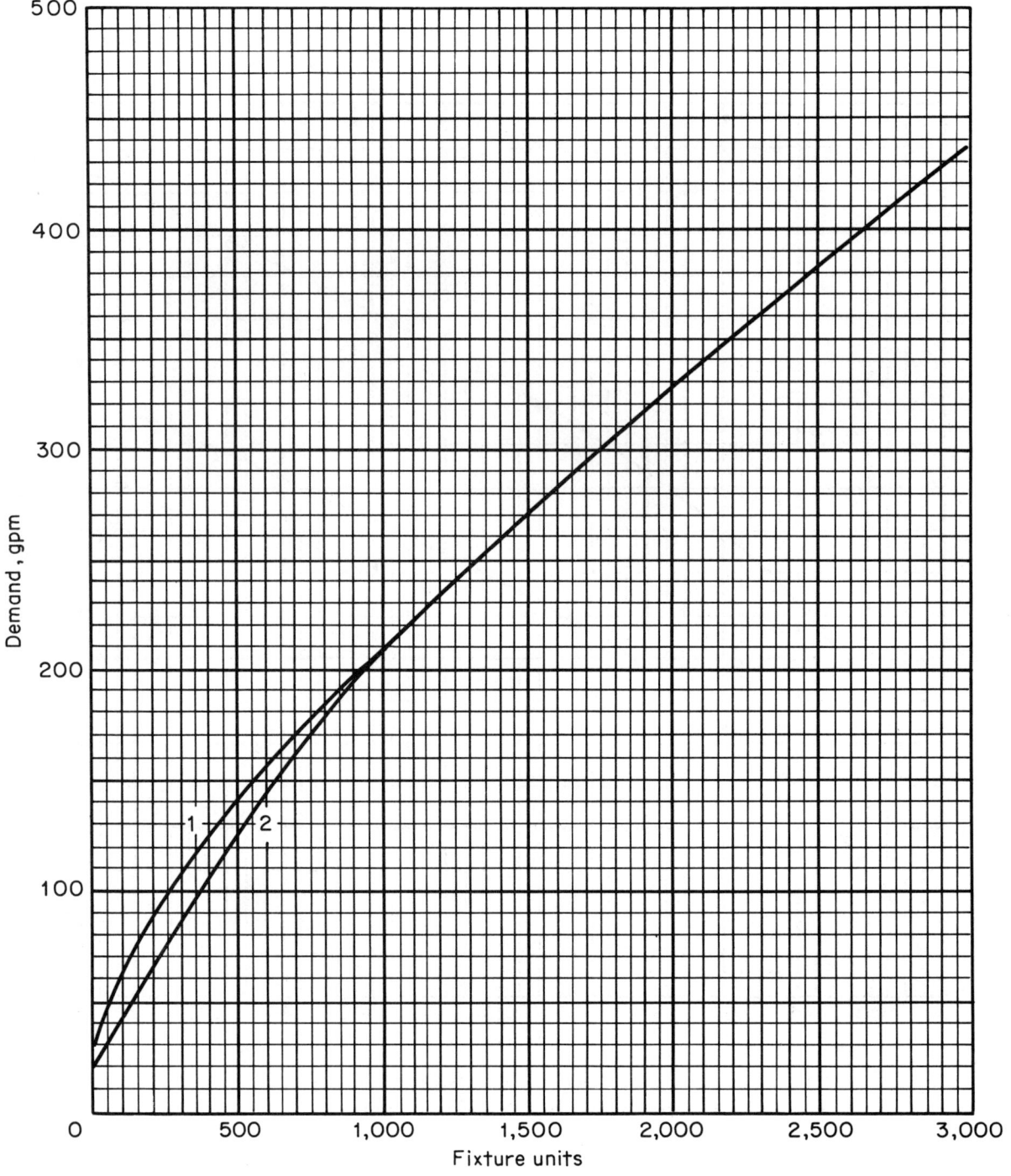

Curve 1 for system with flush valves
Curve 2 for system with flush tanks

Fig. 4-2. Hunter curves (up to 3,000 fixture units).

From Engineering Manual, Part V, Chapter 4, March 1946 by the Department of the Army, USA.

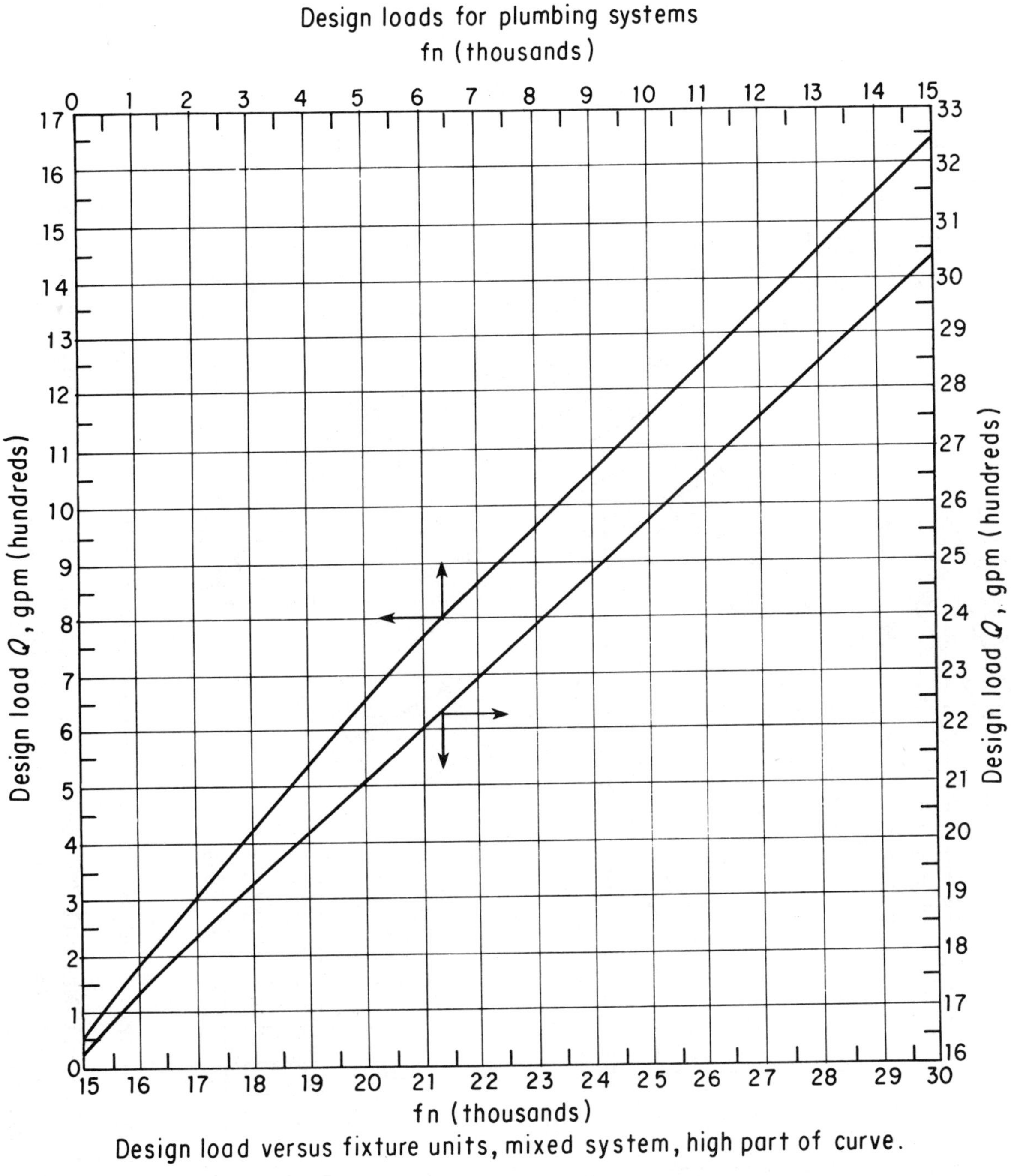

Design loads for plumbing systems
fn (thousands)

Design load versus fixture units, mixed system, high part of curve.

Fig. 4-3. Expanded Hunter curves (up to 30,000 fixture units).

From National Plumbing Code Handbook by Vincent T. Manas. Copyright (c) 1957 by McGraw-Hill, Inc. Used with permission of McGraw-Hill Book Company.

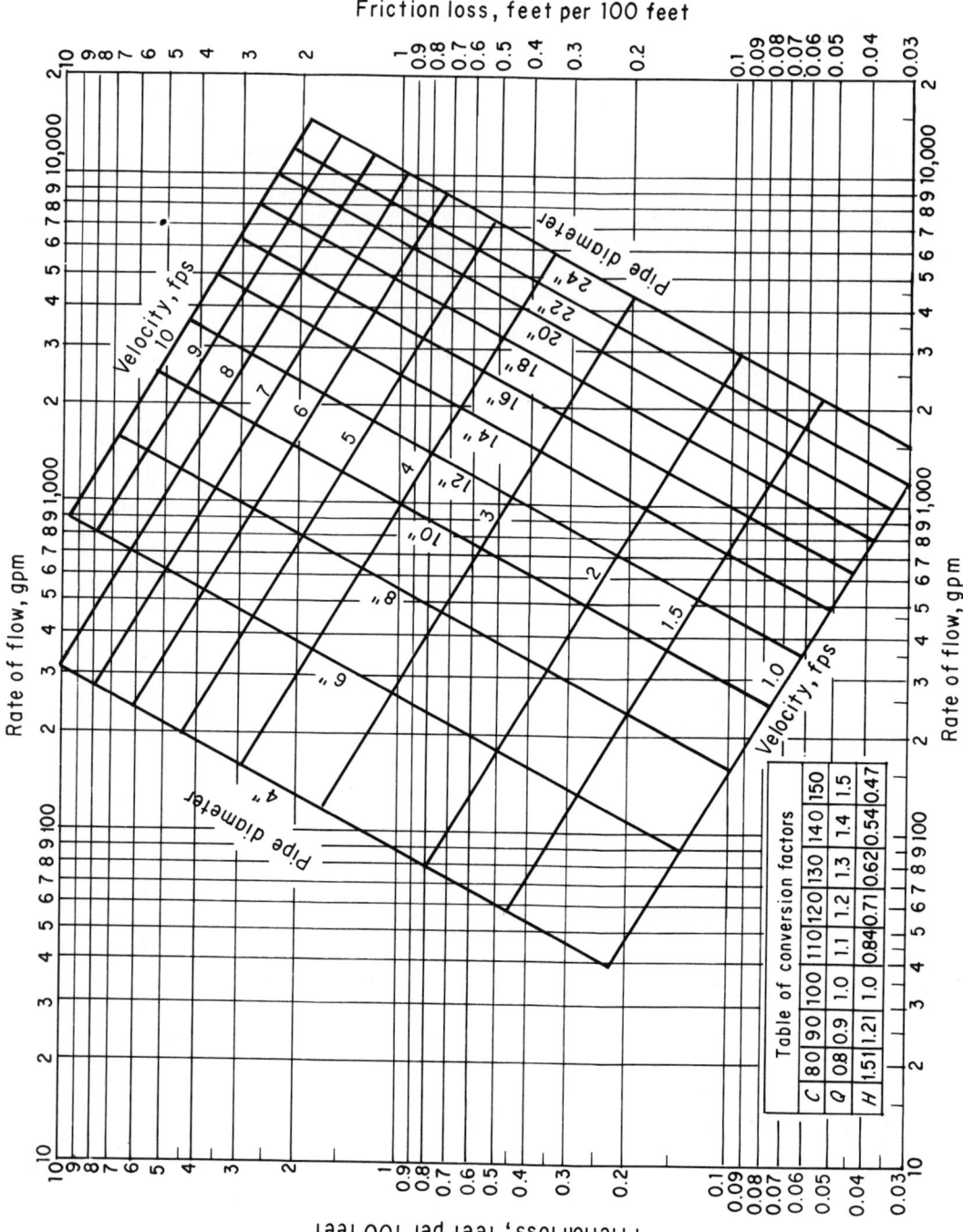

Fig. 4-4. Hazen and Williams pipe flow chart (friction table).

From Technical Manual TM 5-814-1, Sanitary Engineering, Sanitary and Industrial Waste Sewers, August 1966, by the Department of the Army, USA.

Table 4-5 Load values assigned to fixtures

Fixture	Occupancy	Type of supply control	Load values, in water-supply fixture units		
			Cold	Hot	Total
Water closet	Public	Flush valve	10.0	...	10.0
Water closet	Public	Flush tank	5.0	...	5.0
Urinal	Public	1-in flush valve	10.0	...	10.0
Urinal	Public	3/4-in flush valve	5.0	...	5.0
Urinal	Public	Flush tank	3.0	...	3.0
Lavatory	Public	Faucet	1.5	1.5	2.0
Bathtub	Public	Faucet	3.0	3.0	4.0
Shower head	Public	Mixing valve	3.0	3.0	4.0
Service sink	Offices, etc.	Faucet	2.25	2.25	3.0
Kitchen sink	Hotel, restaurant	Faucet	3.0	3.0	4.0
Drinking fountain	Offices, etc.	3/8-in valve	0.25	...	0.25
Water closet	Private	Flush valve	6.0	...	6.0
Water closet	Private	Flush tank	3.0	...	3.0
Lavatory	Private	Faucet	0.75	0.75	1.0
Bathtub	Private	Faucet	1.5	1.5	2.0
Shower stall	Private	Mixing valve	1.5	1.5	2.0
Kitchen sink	Private	Faucet	1.5	1.5	2.0
Laundry trays (1 to 3)	Private	Faucet	2.25	2.25	3.0
Combination fixture	Private	Faucet	2.25	2.25	3.0
Dishwashing machine	Private	Automatic	1.0	1.0	1.0
Laundry machine (8 lb)	Private	Automatic	1.5	1.5	2.0
Laundry machine (8 lb)	Public or general	Automatic	2.25	2.25	3.0
Laundry machine (16 lb)	Public or general	Automatic	3.0	3.0	4.0

Note: For fixtures not listed, loads should be assumed by comparing the fixture to one listed using water in similar quantities and at similar rates. The assigned loads for fixtures with both hot and cold water supplies are given for separate hot- and cold-water loads and for total load, the separate hot-and cold-water loads being three-fourths the total load for the fixture in each case.

Source: National Standard Plumbing Code, Appendix B, Table B.5.2.

Table 4-6 Estimating demand

Supply systems predominantly for flush tanks		Supply systems predominantly for flush valves	
Load (water-supply fixture units)	Demand gpm	Load (water-supply fixture units)	Demand gpm
6	5		
8	6.5		
10	8	10	27
12	9.2	12	28.6
14	10.4	14	30.2
16	11.6	16	31.8
18	12.8	18	33.4
20	14	20	35
25	17	25	38
30	20	30	41
35	22.5	35	43.8
40	24.8	40	46.5
45	27	45	49
50	29	50	51.5
60	32	60	55
70	35	70	58.5
80	38	80	62
90	41	90	64.8
100	43.5	100	67.5
120	48	120	72.5
140	52.5	140	77.5
160	57	160	82.5
180	61	180	87
200	65	200	91.5
225	70	225	97
250	75	250	101
275	80	275	105.5
300	85	300	110
400	105	400	126
500	125	500	142
750	170	750	178
1,000	208	1,000	208
1,250	240	1,250	240
1,500	267	1,500	267
1,750	294	1,750	294
2,000	321	2,000	321
2,250	348	2,250	348
2,500	375	2,500	375
2,750	402	2,750	402
3,000	432	3,000	432
4,000	525	4,000	525
5,000	593	5,000	593
6,000	643	6,000	643
7,000	685	7,000	685
8,000	718	8,000	718
9,000	745	9,000	745
10,000	769	10,000	769

Source: National Standard Plumbing Code, Chapter 10, Table 10.13.2.B.

Table 4-7 Hospital water factors

FU	Hunter, gpm	percent	Adjusted, gpm	Minimum, gpm
Up to 400	125	100	125	
401 - 600	155	90	140	130
601 - 1,200	235	77	180	145
1,201 - 1,500	270	74	200	185
1,501 - 2,000	330	70	230	205
2,001 - 2,500	385	69	265	235
2,501 - 3,000	435	68	295	270
3,001 - 4,000	560	65	365	300
4,001 - 5,000	675	64	430	370
5,001 - 6,000	775	63	490	435
6,001 - 8,000	975	62	600	495
8,001 - 10,000	1,175	61	720	605
10,001 - 13,000	1,460	60	875	725

Table 4-8 Office buildings, schools, and apartment water factors*

FD	Hunter, gpm	percent	Adjusted, gpm	Minimum, gpm
Up to 400	125	100	125	
401 - 600	155	87	135	130
601 - 900	195	75	145	140
901 - 1,200	235	64	150	150
1,201 - 1,500	270	63	170	155
1,501 - 2,000	330	61	200	175
2,001 - 2,500	385	60	230	205
2,501 - 3,000	435	59	255	235
3,001 - 4,000	550	58	320	260
4,001 - 5,000	675	56	380	325
5,001 - 6,000	775	56	435	385
6,001 - 7,000	875	56	490	440
7,001 - 8,000	975	55	540	495

*Add gym showers and laboratory outlets separately.

Table 4-9 Water-pipe sizes: Number of 1/2 connections allowed

Pipe size, in	Average demand	100 percent demand
1/2	1	1
3/4	4	3
1	10	6
1-1/4	20	12
1-1/2	30	20
2	50	35
2-1/2	90	60
3	125	85
4	225	150

Table 4-10 Water-pipe sizes: Flush valves*

Pipe size, in	Number of 1 in flush valves allowed
1-1/4	1
1-1/2	2-4
2	5-12
2-1/2	13-25
3	26-40
4	41-100

*Figure that two 3/4-in flush valves equals one 1-in flush valve but can be served by 1-in pipe. Water-pipe sizing must be tempered by consideration of demand factor, available pressure, and length of run.

Table 4-11 Water-pipe sizes (1/2-gpm lavatory faucets)[a]

Pipe size, in	Average use	100 percent use
3/8	2	2
1/2	5	4
3/4	12	10
1	25	20

[a] In general pipe sizing, allow 1/2 FU for 1/2-gpm lavatory faucet. Water-pipe sizing must be tempered by consideration of demand use, available pressure, and length of run.

2. Street-pressure systems.

   a. Principles of design.

   (1) Where the street pressure is sufficient to supply all the building requirements, provide the building with a street-pressure system.
   (2) Consider street pressure adequate when the street pressure (discounted 5 psi for future unknowns) is equal to or greater than the sum of the required pressure at the highest fixture or equipment, the height loss, the meter loss, and the friction loss to the highest farthest fixture or equipment.
   (3) Cross-connect street-pressure systems of adjoining buildings to provide a second service.
   (4) Street-pressure systems may be used to supply the lower floors of buildings requiring boosted pressure systems where they may be economically justified.

3. Boosted-pressure systems.

   a. Principles of design.

   (1) Provide boosted pressure systems when the street pressure cannot maintain the required pressure to all fixtures and equipment at peak flow 100 percent of the time.
   (2) Boosted pressure systems should normally be either a gravity (house) tank or constant-pressure booster pumps, or for special conditions, a hydropneumatic tank.

      (a) A study for each project should be made to determine which system is the most practical.
      (b) If a constant-pressure booster system is selected, a further analysis should be made to determine if it should be a constant speed or a variable-speed system.

(3) Booster pump systems:

    (a) Generally small systems (up to 40 hp), except where there is over 20-psi fluctuaction in suction pressure, should be constant-speed systems.
    (b) Small systems with horizontal-shaft centrifugal pumps can be variable-speed fluid-coupling drive types in sizes 5 hp and less that are air-cooled. Larger sizes are water-cooled and therefore violate water-conservation design.
    (c) For systems 50 hp and larger and where there is over 20-psi fluctuation in suction pressure consider SCR (electrical) variable-speed booster systems.
    (d) Pressure control for SCR (electrical) variable-speed systems should preferably be located at the far end of the system.

(4) Where an intermediate tank is required, a study should be made to determine which is the most economical--pumping all the water to the top tank and filling the intermediate tank from it or using separate sets of house pumps to fill each tank. Generally, if the building does not materially widen out at the base, one set of house pumps is the most practical.

(5) Use of three pumps rather than two pumps should be studied. Three pumps are desirable where some loads are seasonal (cooling-tower makeup) and give smaller installed capacity (including standby), which under some codes may save the use of a suction tank.

(6) Where future loads are involved, design the ultimate system with only the initial requirements installed.

(7) Where a house tank is required by code for the fire standpipe system, generally, it rather than a constant-pressure system should also be used for the domestic water.

(8) When the height of the building produces excessive pressures, the building should be zoned with master pressure-reducing stations for lower zones.

   (a) Zoning must be given careful study. Zones require additional distribution piping and hot-water heaters. However, many branch pressure-reducing valves are a maintenance headache.
   (b) Maximum-height zones should generally be used with a few floors of each zone having branch pressure-reducing valves, if necessary to save a zone.
   (c) Where an intermediate tank is required for the fire standpipe system, it should generally also be used for the domestic water, thus eliminating one pressure-reducing valve zone and reducing the pressure in the piping to lower pressure-reducing valve zones.

(9) Gravity house tank systems should use downfeed risers. Constant-pressure systems should use downfeed risers for all zones except the top.

(10) When a boosted pressure system is required for a building, individual conditions should decide whether it is economically more practical to supply the entire building on boosted pressure or part on boosted pressure and part on street pressure. Factors to be considered include the number of floors that can be fed from street pressure, the size of the load on these floors, and the similarity or dissimilarity with the layout of the floors above. In the latter case, it is usually best to feed continuing similar floors by boosted pressure.

(11) Where a gravity house tank cannot be placed high enough to provide all the pressure requirements of fixtures and equipment on the topmost floors, a small constant-pressure system or hydropneumatic pressure tank system pumping from the house tank should be provided to serve these fixtures and equipment.

(12) Where the code allows it and the owner agrees to accept the insurance penalty, constant-speed constant-pressure booster pumps can double as fire pumps if they have the required

capacity.
(13) Provide house pumps and constant-pressure booster pumps with a low-suction-pressure switch to shut them down in the event of loss of water supply.
(14) Where required by code, where extensive on-site storage is required because of the lack of a second service, or where water comes from an on-site water-treatment plant, provide a suction tank.

   (a) In New York City projects, where the installed house- or booster-pump capacity exceeds 400 gpm, negotiations with the department of water supply are necessary to eliminate the need of a suction tank. Before you talk to the department, obtain and present a hydrant test; because without it, they will not discuss the matter.

(15) To prevent the pump suction from robbing the gravity feed, and to prevent the pull of lower-zone demands from robbing the top-zone gravity feed, use separate connections to the house tank and separate piping for each of the following:

   (a) Cold water to top gravity tank zone.
   (b) Water to cooling tower and/or hydropneumatic pressure tank pumps.
   (c) Cold-water to hot-water heaters, lower pressure-reducing valve zones, and lower gravity tanks.

(16) The sizing of the branches from the mains to the risers and the risers down to the first branch should be generous, preferably one size larger than the riser below, unless the pipe size has just increased below.
(17) Distribution should preferably be at the ceiling below the top story served; but if necessary, it can be at the ceiling of the top story served.

b. Equipment:

   (1) Constant-pressure booster pumps.

(a) The gpm capacity of the pumps should be the same as that calculated for sizing the water piping of the system.
(b) In a two-pump installation, each pump should be full size. In a three-pump installation, each pump should be half size, or two should be large and one small. The capacity of one large pump and the small pump should equal the peak load. It is desirable to have a small pump to carry the load in off-peak hours. The best solution is a half-size pump, if it is small enough. If not, a small pump is required (size depending on project requirements).
(c) Calculate the pump head by adding the pressure required at the farthest and/or topmost fixture or equipment (converted from psi to feet), the lift from the pump centerline elevation to the topmost fixture or equipment, the friction at peak flow in the piping from the pump to the topmost fixture or equipment (pipe-sizing calculation), plus 10 ft for controls, minus the minimum available suction pressure at the pump centerline elevation (converted from psi to feet).
(d) Multistage diffuser pumps (can pumps) have the best curves for this service and the best efficiencies; however, in low heads and small sizes, they are the most expensive.
(e) Horizontally split case pumps must be considered in low head ranges. Vertically split case pumps should be considered only in small sizes.
(f) Factory-assembled package units should be considered.
(g) Except for buildings with 24-hour, 365-day continuous demand, provide off-hour operation, either on-off with time delay or small pressure tank.
(h) For variable-speed pumps, the discharge check valve must be spring type.
(i) With variable-speed squirrel-cage induction motors, the peak running current at reduced speed far exceeds that for the same hp constant-speed motors. This peak current must be obtained and reviewed

with the electrical project engineer in order that he may correctly size his wiring.

(2) House (tank-fill) pumps.

   (a) Pumps should be duplex (each full size) or triplex (each half size), whichever best fits the building load.
   (b) For house tanks, calculate the gpm capacity of the pumps by multiplying the number of fixtures (not fixture units) by the factors in Table 4-12.
   (c) Calculate the pump head by adding the lift from the pump centerline elevation to 1 ft above the top of the gravity tank and the friction in the pump discharge from the pump to the tank, minus the minimum available suction pressure at the pump centerline elevation (converted to feet). Where float valves are provided, the loss through the float valve must be added to the head of the pump.
   (d) Pumps should be horizontally split case type, multistage where required, or multistage diffuser pumps.
   (e) In sizing the house pump discharge, in sizing and selecting the house pumps, and in designing the house tank, consider future building expansion, especially if it is to be upward.

   (i) The pump discharge should be sized for the ultimate.
   (ii) In the future, either a third pump should be added or the pumps and motors selected so as to allow the use of a larger-diameter impeller in the future.
   (iii) The house tank should be two tanks instead of a two-compartment tank to facilitate raising the tanks in the future, plus adding a third tank in the future if necessary.

   (f) For tanks on a tower, if a well pump is the supply, it should be sized to fill the domestic capacity (including cooling-tower makeup) of the tank in a maximum of 16 hours. If water is from any other source, size it to fill the domestic capacity (including cooling-tower makeup) in a maximum of 20 hours.

(g) The house pump discharge check valves must be spring-loaded type instead of the swing type used elsewhere. Where the pump draws from a suction tank on the same level as the pump, a similar spring-loaded-type check valve must be provided in each pump suction line.

(h) The discharge of the house pumps must be provided with float valves or the pumps provided with an emergency high-water-limit switch. Where the possibility of night creep (increased street pressure) exists, or surging of street pressure could feed water into the gravity tank, provide float valves for each compartment of the gravity tank. In other cases, simply use two-pole high-water-alarm switches with 1-pole acting as high-limit switch.

(3) Gravity tanks.

(a) The house tanks should have a domestic storage of 40 times the pump's capacity for pumps 100 gpm and less, and 30 times the pump's capacity for larger pumps with 4,000 gallons as minimum. To this quantity, add the reserve for the fire stand-pipe system and the sprinklers as required by code and/or underwriters.

(b) Where house pumps are supplying cooling-tower makeup as well as domestic water requirements, calculate the tanks domestic storage as above, using the domestic portion of the pump's capacity plus 15 times the pump's cooling-tower makeup capacity.

(c) Where a top tank is feeding an intermediate tank, calculate its domestic storage using 30 times the capacity that the pump would have if only feeding the top tank, plus 15 times the capacity the pump would have if only feeding the

intermediate tank, plus the cooling-tower requirements as above.
(d) An intermediate tank should have a domestic storage capacity of 30 times the capacity the pump would have if only feeding the intermediate tank, plus the reserve for the fire standpipe system and the sprinklers as required by code and/or underwriters.
(e) A tank located outside on a tower should have a domestic storage of one day's domestic and cooling-tower requirements plus a fire reserve as required by code and/or underwriters.
(f) House tanks should be rectangular two-compartment steel with outside bracing and cover.

    (i) Tank should rest on 18-in dunnage beams running the short way and spaced approximately on 24-in centers.
    (ii) If the tank room is waterproofed, provide the tank with a condensation gutter piped to spill over a floor drain.
    (iii) If the room is not waterproofed, set the dunnage beams on a steel pan with a piped-up drain in it.
    (iv) Provide the tank with a main drain on the side piped together with the overflow to spill over the roof.
    (v) Provide a 1-1/4-in drain in the bottom in each compartment, directly over a floor drain or pan drain.
    (vi) Provide tank with high- and low-water alarms.
    (vii) See Figure 4-5 for details of tank construction.
    (viii) Allow a minimum of 30 in from the underside of the slab above to the top of the tank for controls, inlet, and access into the tank. Allow a minimum of 2-ft clearance behind and at ends of tank for painting.

>   Allow approximately 8 ft in
>   front of the tank for pipe
>   headers and aisle.
>   (ix) Where the bottom of the tank is
>   more than 24 in above the floor,
>   have structural engineer pro-
>   vide necessary supports for the
>   dunnage beams.

(g) To calculate the net capacity of a house tank, subtract 18 in from the tank height for mud bottom, controls, and overflow. The minimum height of the tank should be 8 ft. Tanks over 12 ft high get overly expensive due to increased bracing requirements.

(h) To obtain the approximate weight of a steel house tank and the water in it for the structural engineer, multiply the gross capacity of the tank in gallons by 11 lb/gal.

(i) While inside tanks can be round wooden tanks, wood is definitely a second choice inside, although a first choice outside. For details of construction of wooden tanks, see NFPA Standard No. 22.

(j) Tanks on towers should be elipsoidal round steel or round wood depending on the sizes, economics, and esthetics. For details of construction of tanks and towers, see NFPA Standard No. 22.

(k) In freezing climates, all outside steel tanks should be heated, including outside wooden tanks, with the possible exception of those for hospitals and large apartment buildings where water is used 24 hours a day.

(l) Suction tanks should be similar construction to house tanks and sized as required by code, and/or underwriters; or, in the absence of requirements, with a net capacity of at least 15 times the maximum pumping rate from the tank. (Use larger tank if on-site storage is desired.) Tanks should be filled through a float valve and a flow-limiting orifice and provided with an overflow, drain, and high- and low-water alarms.

(4) Hydropneumatic pressure tank pumps.

  (a) The pumps should be duplex (each full size) or triplex (each half size), whichever best fits the load.
  (b) Calculate the gpm capacity of the pumps by multiplying the number of fixtures (not fixture units) by the factors in Table 4-12.
  (c) Calculate the pump head by adding the pressure required at the farthest and/or topmost fixture or equipment (converted from psi to feet), the lift from the pump centerline elevation to the topmost fixture or equipment (a negative number if pump is above the fixture or equipment), the friction at peak flow in the piping from the pump to the topmost fixture or equipment (pipe-sizing calculation), plus an operating pressure differential (converted from psi to feet) (usually 20 psi), minus the minimum available suction at the pump centerline elevation pressure (converted from psi to feet).
  (d) Provide the pump discharge header with a bypass relief valve and sight-flow indicator back into the suction header.
  (e) Provide pressure relief on the hydropneumatic tank pump discharge piping as well as on the tank.
  (f) Control the pumps by a system controller, to properly cycle the pumps and maintain the proper air pressure in the tank.
  (g) The pumps should be horizontally split case type except that vertically split case pumps may be used in small sizes.
  (h) Select the pumps at rated gpm at the starting (turn-on) head. It must have a curve steep enough to reach the stopping head (starting head plus the operating cycle) at reduced capacity. This is often a difficult selection to find, because the 20-psi operating cycle requires a pump with a very steep curve. It may often be necessary to select the rated gpm with a head greater than calculated, in order to be able to get a curve covering both points.

(5) Hydropneumatic pressure tanks.

   (a) The tanks should be ASME stamped-steel pressure tanks, either vertical or horizontal, depending on the size and the available space.

      (i) Provide single tanks with a bypass.
      (ii) Provide the tanks with a relief valve and a pressure gauge.

   (b) Size the tanks as follows: For pump requirements up to 100 gpm, provide a tank 30 times the pump requirements. For larger pump requirements, provide a tank 25 times the pump requirements (minimum capacity, 3,000 gal.).

   (c) Select the air-water ratio carefully for the pressures involved, to get a maximum amount of water out of the tank per cycle and still leave at least 10 percent in the tank at pump start. Note that in horizontal tanks, the depth of the water in the tank in relation to the tank diameter is not the same as the air-water ratio as it is in a vertical tank.

   (d) In tanks feed from well pumps, an excess of air usually accumulates in the tank. This must be bled off through a solenoid valve operated by the system controller. In all other cases, air must be periodically added (as it is absorbed by the water), either by an air compressor or a solenoid valve from a building compressed-air system, operated by the system controller.

   (e) Small vertical 12- to 36-in diameter tanks (not ASME construction and maximum 75 psi wwp) can be provided with a floating diaphragm which eliminates the need for an air compressor.

   (f) Where the 20-psi operating cycle plus the variable friction loss in the piping would give an objectionable pressure variation in the system, a constant-pressure valve (master pressure-reducing valve) should be provided after the tank. Where this pressure-reducing valve is provided, allow an additional 5-psi in the pump head.

(g) To obtain water capacity per cycle: Air expands in proportion to reduction in absoblute pressure (absolute pressure equals gauge pressure plus 14.7 lb/in). Calculate the expansion of the air in the tank per cycle (difference in pressure between pump cutout and cutin). Then subtract the original volume of air. The result gives water pushed out by expanding air per cycle.
(h) See Figure 4-6 for water depths and Table 4-13 for withdrawal quantities.
(i) Refer to Section 0, 5, c for data on air compressors.
(j) Air compressors for domestic hydropneumatic tanks are required to build up pressure in the tank whenever it is drained and refilled, and to make up air that is absorbed by the water.

   (i) The size of the compressor should be the cfm required to build up the pressure in the tank in the maximum practical allowable time considering the type of project involved and the function of the tank.
   (ii) The air absorbed by the water is small and does not affect the compressor sizing.

Table 4-12 Table of factors for sizing house pumps

No. of fixtures	gpm per fixture	Minimum pump capacity, gpm
Apartments (apartment hotels)		
1-25	0.6	10
26-50	0.5	15
51-100	0.35	30
101-200	0.3	40
201-400	0.28	65
401-800	0.25	120
801-up	0.24	210
Hotels and clubs		
1- 50	0.65	25
51- 100	0.55	35
101- 200	0.45	60
201- 400	0.35	100
401- 800	0.275	150
801-1,200	0.25	225
1,201-up	0.2	300
Hospitals		
1-50	1.0	25
51-100	0.8	55
101-200	0.6	85
201-400	0.5	125
401-up	0.4	210
Schools		
1-10	1.5	10
11-25	1.0	15
26-50	0.8	30
51-100	0.6	45
101-200	0.5	65
201-up	0.4	110

(continued on next page)

Table 4-12 (Continued)

No. of fixtures	gpm per fixture	Minimum pump capacity, gpm
Office buildings		
1- 25	1.25	25
26- 50	0.9	35
51- 100	0.7	50
101- 150	0.65	75
151- 250	0.55	100
251- 500	0.45	140
501- 750	0.35	230
751-1,000	0.3	270
1,001-up	0.275	310
Industrial buildings		
1-25	1.5	25
26-50	1.0	40
51-100	0.75	60
101-150	0.7	80
151-250	0.65	110
251-up	0.6	165

Notes: Where fire standpipe or sprinklers are involved, minimum pump capacity is 65 gpm. The figures do not include any water for process work. For laundries in the building, add 10 percent to pump capacity. If majority of occupants are women, add 15 percent to pump capacity.

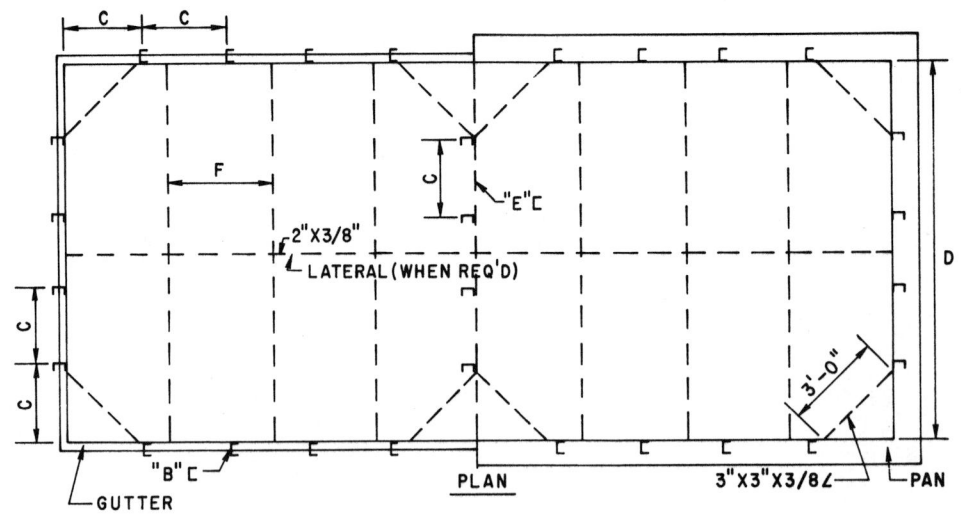

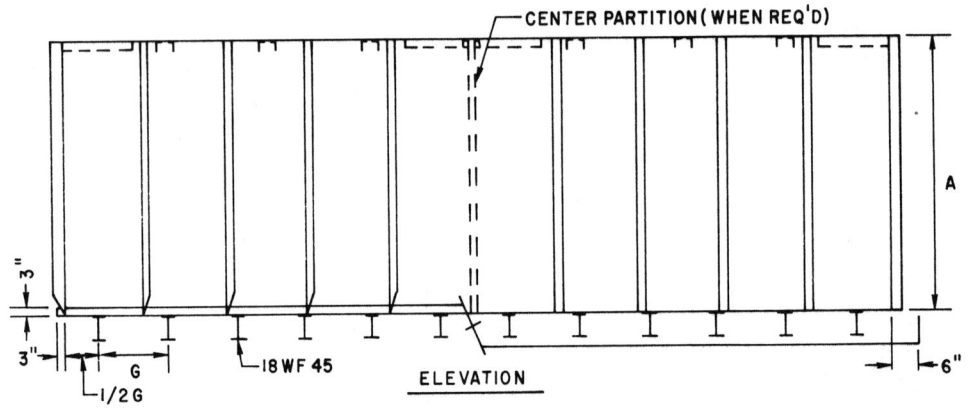

	Channel Stiffeners				Cover Support			
A	B			Maximum spacing C	D	E Channel	Maximum spacing F	Number lateral
Up to 6'0" deep	5 C	6.7#		36 " centers	Up to 7'0" wide	4" 5.4#	48"	-
7'0"	6 C	8.2#		34½"	8'0"	4" 5.4#	45"	-
8'0"	6 C	8.2#		30½"	9'0"	4" 5.4#	42"	-
9'0"	7 C	9.8#		28½"	10'0"	4" 5.4#	42"	1
10'0"	8 C	11.5#		27"	11'0"	4" 5.4#	42"	1
11'0"	8 WF	13	#	25½"	12'0"	6" 8.2#	45"	1
12'0"	8 WF	15	#	24½"	13'0"	6" 8.2#	45"	2
13'0"	8 WF	17	#	23¾"	14'0"	6" 8.2#	45"	2
14'0"	8 WF	20	#	22½"	15'0"	6" 8.2#	45"	3
15'0"	10 WF	21	#	21½"				
16'0"	10 WF	25	#	20-3/4"				
17'0"	10 WF	29	#	20 "				
18'0"	10 WF	33	#	19½"				
19'0"	12 WF	31	#	19 "				
20'0"	12 WF	36	#	18½"				

Dunnage Beams

G: 2'0" Normal, 2'3" Maximum

Plates
Side, bottom, and center: 3/8"
Cover: 3/16"
Pan: 1/4"

Fig. 4-5. Steel house tank construction.

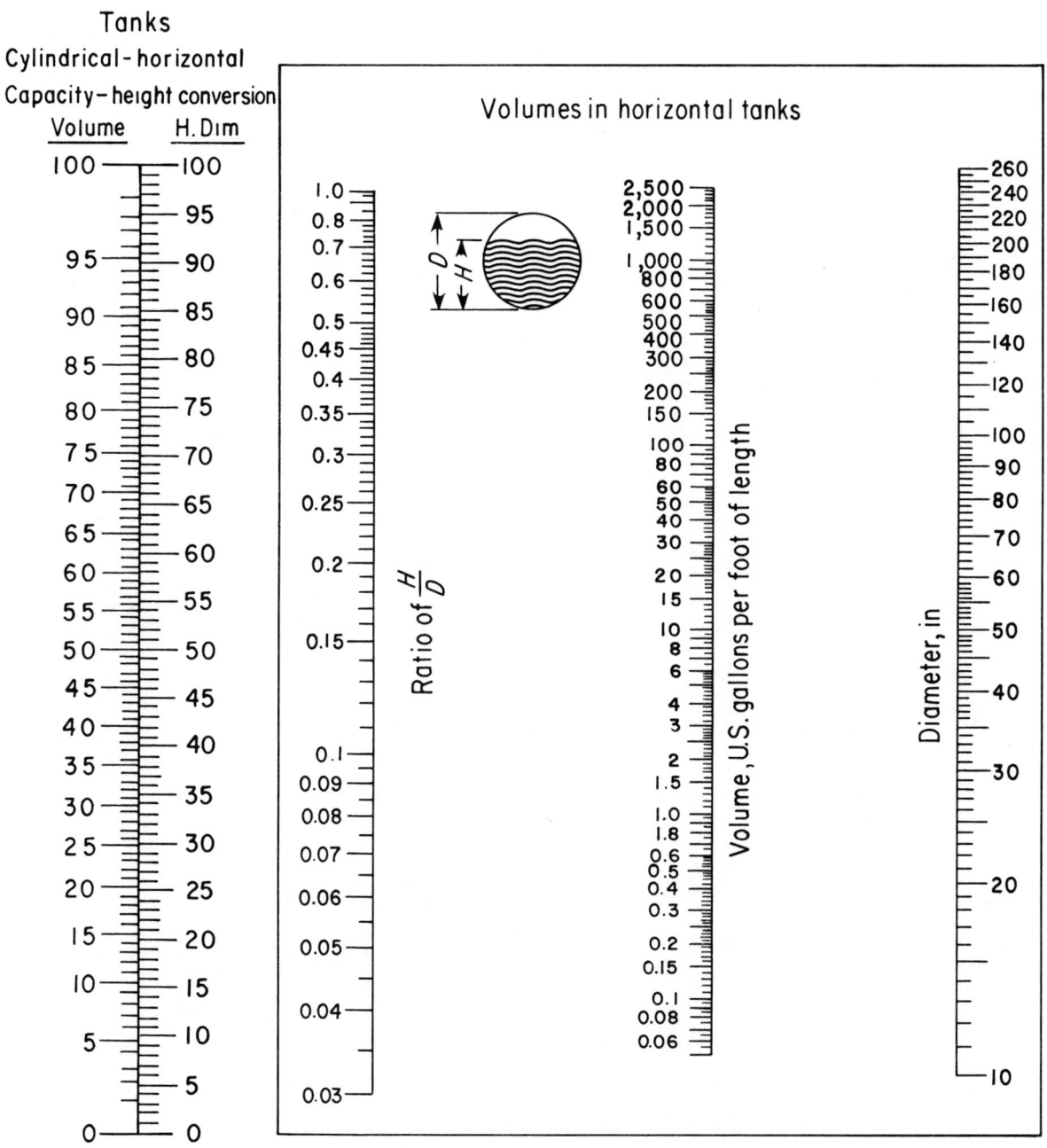

Fig. 4-6. Volumes in horizontal hydropneumatic pressure tanks.

Reprinted from Peerless Pump Bulletin B-579, 1955.

Table 4-13

% OF VOLUME WITHDRAWN BETWEEN STATED PRESSURES WITH TANK FILLED TO 33% OF CAPACITY. TANK EMPTY WHEN 33% OR MORE IS WITHDRAWN.

LBS. GAUGE PRESSURE (LOW OR CUT IN) \ HIGH OR CUT OUT PRESSURE	25	30	35	40	45	50	55	60	65	70	75	80	85	90	95	100	110	120	130	140	150
100																	6	11	17	23	29
95															3	9	15	21	27	33
90														3	6	13	19	25	32	33+
85													3	7	10	17	23	30	33+	
80												3	7	11	14	21	28	33+		
75											4	7	11	15	19	26	33			
70										4	8	12	16	19	23	31	33+			
65									4	8	13	17	21	25	29	33+				
60								5	9	13	18	22	27	31	33+					
55							5	9	14	19	24	29	33	33+						
50						5	10	15	21	26	31	33+	33+							
45					5	11	17	22	28	33	33+	33+								
40				6	12	18	24	31	33+	33+										
35			7	13	20	27	33	33+												
30		7	15	22	30	33+	33+													
25	8	17	25	33	33+															
20	9	19	29	33+	33+															

% OF VOLUME WITHDRAWN BETWEEN STATED PRESSURES WITH TANK FILLED TO 40% OF CAPACITY. TANK EMPTY WHEN 40% OR MORE IS WITHDRAWN.

LBS. GAUGE PRESSURE (LOW OR CUT IN) \ HIGH OR CUT OUT PRESSURE	25	30	35	40	45	50	55	60	65	70	75	80	85	90	95	100	110	120	130	140	150
100																	5	10	16	21	26
95																2	8	14	19	25	30
90															3	5	11	17	23	29	34
85														3	6	9	15	21	27	33	39
80													3	6	10	13	19	25	32	38	40+
75												4	7	10	13	17	23	30	37	40+	
70											4	7	11	14	17	21	28	35	40+		
65										4	7	11	15	19	23	26	34	40+			
60									4	8	12	16	20	24	28	32	40+				
55								4	8	13	17	22	26	30	34	38	40+				
50							5	9	14	19	23	28	32	37	40+	40+					
45						5	10	15	20	25	30	35	40+	40+							
40					5	11	16	22	28	33	38	40+									
35				6	12	18	24	30	36	40+	40+										
30			7	13	20	27	34	40+	40+												
25		7	15	23	30	38	40+														
20	8	17	26	35	40+	40+															

% OF VOLUME WITHDRAWN BETWEEN STATED PRESSURES WITH TANK FILLED TO 50% OF CAPACITY. TANK EMPTY WHEN 50% OR MORE IS WITHDRAWN.

LBS. GAUGE PRESSURE (LOW OR CUT IN) \ HIGH OR CUT OUT PRESSURE	25	30	35	40	45	50	55	60	65	70	75	80	85	90	95	100	110	120	130	140	150
100																	4	8	13	17	21
95																2	7	11	16	20	25
90															2	4	9	14	19	24	28
85														2	5	7	12	17	22	27	32
80													2	5	8	10	16	21	26	31	37
75												3	5	8	11	14	19	25	30	36	42
70											3	6	9	12	14	17	23	29	35	41	47
65										3	6	9	12	15	19	22	28	34	40	47	50+
60									3	6	10	13	16	20	23	26	33	40	47	50+	
55								3	7	10	14	18	21	25	28	32	39	46	50+		
50							4	7	11	15	19	23	27	31	34	38	46	50+			
45						4	8	12	16	21	25	29	33	37	42	46	50+				
40					4	9	13	18	23	27	32	36	41	45	50+	50+					
35				5	10	15	20	25	30	35	40	45	50+	50+							
30			5	11	16	22	28	33	39	44	50+	50+									
25		6	12	19	25	31	37	44	50+	50+											
20	7	14	21	29	36	43	50+	50+													

Calculations made on basis of absolute pressures at sea level (gauge pressure plus 14.7 psi). Reprinted from Power, Sept. 1948. Copyright McGraw-Hill, Inc., 1975. All rights reserved.

4. Hot-water systems.

   a. Principles of design.

   (1) Provide complete circulated hot-water-supply system in each building for the street pressure zone and each boosted pressure zone.
   (2) Provide major kitchen facilities with a separate hot-water system (140°F) from the remainder of the zone facilities.
   (3) Provide laundries with commercial-type washers with a separate hot-water system (180°F), if large, including waste heat reclaimer with rotating screen and circulating pump, and storage-type heaters with full temperature rise.
   (4) Unless otherwise required for a special reason, hot-water heaters should be controlled-heat semi-instantaneous type.
   (5) Where electricity is the energy source, hot-water heaters should be storage type to minimize the kilowatt heating requirements.
   (6) For hospitals, major kitchens, major laundries, and other critical services, heaters should be duplex (each half size) and single elsewhere.
   (7) Whenever waste-heat Btu's are available, provide a preheater to take advantage of them. One preheater can serve more than one set of hot-water heaters, provided they are in the same location and at the same pressure.
   (8) Booster hot-water heaters for 180°F hot water for kitchen washing machines will normally be provided with the washers.
   (9) Where the building is generally provided with 95° tempered hot water and a few isolated fixtures require a higher temperature, provide a small electric booster heater at those fixtures.
   (10) Where the building is generally provided with 95° tempered hot water and a significant number of fixtures require a higher temperature, consider the following in order of preference:

   (a) Make the building system 120°F and use tempering valves where required to obtain the 95°F water.
   (b) Provide a separate system for each.

(c) Abandon the 95°F water and use normal 120°F instead.

(11) Provide all hot-water systems with pumped circulation piping back to the heater. For hospitals, major kitchens, major laundries, and other critical loads, circulators should be duplex (each full size) and single elsewhere.

(12) Provide hot-water-circulated piping veritcal and horizontal as required to supply all fixtures and equipment requiring hot water. Plan an orderly balanced system with an absolute minimum of subcirculating loops or unreasonably complicated routing of mains.

(a) Interconnect the bottoms of downfeed risers and run back to circulators.
(b) Where practical, make connections to upfeed risers at the ceiling below the top branch and run horizontally, connecting all risers, then down to circulators, rather than a separate circulation line paralleling each riser.
(c) See Table 4-16 for the maximum allowable lengths of dead-leg hot-water branches.

(13) Proper sizing of the hot-water circulating system is essential for the efficient and economical operation of the system. Undersizing of the piping or circulating pump will seriously hamper circulation, and thus adequate-temperature hot water will not be immediately available at all fixtures. See hot-water circulating pumps for sizing procedures.

(14) Provide tempering valves wherever the possibility of overheating the hot water exists, and when piping serves small children. Tempering valves should also be considered on piping serving the aged and infirm.

(15) Provide thermostatic mixing valves wherever accurate temperatures must be delivered. As thermostatic valves are specification rated at 45 psi, they must be selected to give the required flows at the available pressure, then the selected size must be converted to flow rate at 45 psi for specifying.

(16) When hot water is required for hose bibbs, and is not readily available from the regular hot-water system, provide steam-water thermostatic mixing valves. Pressures of steam and water should be relatively equal with the water pressure slightly higher than the steam pressure.

(17) Heaters should preferably be located at the bottom of their system.

b. Equipment.

(1) General.

(a) Controlled-heat semi-instantaneous-type heaters and heaters with pumped reversed-flow internal circulation should be set at the desired delivery temperature as follows: General building systems (110°F water to fixtures), 120°F; general building systems (95°F water to lavatories), 105°F; kitchens (140°F water to fixtures), 145°F; major laundries (180°F water to equipment), 185°F.

(b) Normal storage-type heaters should be set 20°F above the desired delivery temperature to allow for cooling down to desired delivery temperature as storage is used up. (Only approximately 65 to 70 percent of tank volume is usable at or above the desired delivery temperature.)

(c) To save space, where large storage is involved, it is possible to set tank temperature at 200°F with a tempering valve and safety solenoid on the outlet. This will allow a reduction of storage capacity to the point where the 200°F water plus the added cold water equals the required storage.

(d) Steam pressures provided by HVAC for domestic hot-water heaters should be split with half the pressure a loss through the valve and half the pressure a loss through the coils. Where the steam pressure is an increment of 5 psi, the odd psi should be on the coil. Steam pressure on the coil should always be less than the water pressure in the hot-water system.

(e) Where controlled-heat semi-instantaneous-type heaters are used, the following information should be given to the HVAC project engineer: gpm and temperature rise of the heater as specified; the gph calculated by the "storage tank" method for a peak hour at the temperature rise. This will enable the HVAC department to use their judgement in sizing the steam piping and in estimating their boiler load.

(f) In sizing hot-water heaters on a project where there is a definite day and night load, the loads at different times of the day should be carefully analyzed (rational method) and a heater selected to satisfy the greater demand. This is important, as the normal method of heater sizing will provide a needlessly large heater under these circumstances.

(g) Provide all hot-water heaters and preheaters with T&P relief valves (some codes also require an additional separate pressure-relief valve on larger-size tanks). Size the T&P relief valves on their <u>temperature</u> rating (their pressure rating is always higher), using the AGA rating for all gas heaters and for other heaters with 200,000 Btu/h or less input. Use the manufacturer's rating for other heaters with over 200,000 Btu/h input. T&P relief valves have a maximum working pressure of 150 psi; therefore when the system pressure exceeds 150 psi, a pressure-relief valve of required rating and an aquastat and solenoid valve bleed must be provided.

(2) Controlled-heat semi-instantaneous-type hot-water heaters.

(a) The heaters should normally be sized using fixture units and modified "Hunter" curves (Table 4-17 and Figures 4-7 and 4-8).

(1) For 1/2-gpm lavatory faucets, allow 1/2 FU per lavatory.
(2) Duplex heaters should be each half size.

4-68

(b) Use air-operated steam valves whenever control air is available from the HVAC systems.

(c) Provide 18 in behind heaters for control element removal.

(3) Storage-type hot-water heaters.

(a) Size heaters by using Tables 4-18, 4-19, or 4-20.

(b) Where gas or oil is the source of heat, a flue must be provided either connecting to the main building chimney (stack) or running independently through the roof.

(c) Where electric heaters are used, the relationship between the makeup and the storage should be given careful consideration, as electrical makeup builds up a high kilowatt load very quickly.

(i) Within practical limits, makeup should be kept as low as possible, with increased storage to compensate for it.

(ii) Except in residential-type heaters, provide stepped control to minimize the electrical loads, except at full-peak demand periods.

(d) Where electrical load-limiting controls are provided in the building, and theoretically all heating must be done at night, provide a storage of a full day's requirements, and a heating rate of a full day's requirements divided by the number of hours available for heating.

(i) Provide multiple heating elements with one element in each heater active during the normal non-heating hours for "topping off" in the event of an unexpected abnormally high demand.

(ii) Use a heater with a pumped reverse flow internal circulation.

(e) For horizontal heaters, adequate space must be allowed in front for tube pull.

(4) Laundry waste heat reclaimers.

   (a) Laundry waste heat reclaimers should be closed package-type shell and tube heat exchangers to heat-required fresh water in the shell (40 to 105°F) with 120°F waste water (total hot and cold water required for the laundry) in the tubes.

   (i) The unit should include a self-priming nonclog circulating pump for circulation between the waste-water sump pit and the unit, and a rotating screen in the waste-water sump pit.
   (ii) Provide sufficient space on each end of the unit to allow for proper tube cleaning.

   (b) The laundry heater should be a storage type with a useful storage of a peak-hour's supply and a heating rate of a peak-hour's supply at full temperature rise (assuming reclaimer is not functioning). Use a heater with a pumped reverse-flow internal circulation.

   (c) The laundry heater may be a controlled heat semi-instantaneous type sized for the peak gpm demand of the laundry at a temperature rise sufficient to heat a peak-hour's supply at full temperature rise, plus an accumulator tank with a useful storage of a peak-hour's supply and pumped circulation between them.

(5) Preheaters (condensate coolers).

   (a) The preheaters should be vertical storage-type heaters sized to store at least 1 to 1 1/4 times the peak-hour's requirements with a heating element sized to heat a peak-hour's requirements at the maximum temperature rise that the waste heat system can produce, up to the full temperature rise required for the system.

   (b) Provide preheaters with a pumped internal circulation.

- (c) Where the waste heat supply is provided with an air-operated automatic-control valve, have the HVAC designer provide a PE switch in his control system to operate the circulator only when the heating medium is flowing. In other conditions, provide an aquastat for controlling the circulator.
- (d) Where the waste heat system is capable of overheating the water, provide a tempering valve and safety solenoid on the piping to the system to guard against overtemperature water reaching the fixtures.

(6) Hot-water-circulating pumps.

- (a) The hot-water circulating pumps should be all bronze pumps. They can normally be in-line circulators, where available in the size and pressure rating required; otherwise they should be base-mounted compact flexible connected pumps.
- (b) The pumps should normally be provided with H-O-A switches and aquastats.
- (c) Where the hot-water usage does not exist many hours of the week, provide a 7-day time clock to save unnecessary energy use during periods when the system is not in use.
- (d) Where self-contained tempering valves are provided, omit aquastats, because the valve manufacturers prefer constant circulation through their valves.
- (e) For preliminaries and as a check against final calculations, base the capacity of the circulators and/or pumps on comparison of the following two calculations:
    - (i) Allow 1 gpm for every 20 fixtures in the system. Where 1/2-gpm lavatory faucets are used, allow 1 gpm for every 50 fixtures in the system.
    - (ii) Allow 1/2-gpm for each 3/4- and 1-in supply riser. Allow 1 gpm for each 1 1/4- and 1 1/2-in supply riser. Allow 2 gpm for each 2- and 2 1/2-in supply riser. Allow 3 gpm for each 3-in and larger supply riser.

(As used above, riser means a circulated loop either vertical or horizontal.)

(f) Heads shall be generous, normally at least 1 1/2 times that roughly calculated from the measured lengths.

(g) The final calculations should be made as follows:

    (i) Assume 1-in insulation on the circulated piping.

    (ii) Assume a maximum 10°F temperature drop in the entire system for building systems, and a 5°F temperature drop in kitchen and laundry systems.

    (iii) Size the circulated supply piping and then tabulate the measured length of each section.

    (iv) From the Table 4-14, multiply the length by the Btu/(h)(lin ft) heat loss and total the heat loss in the <u>entire</u> supply piping. Tentatively assume the heat loss in the circulating (return) piping to be two-thirds of that in the supply, and total for the whole system.

    (v) Divide the total system heat loss by 5,000 for general building systems and 2,500 for kitchen and laundry systems. This will give the minimum required circulation rate of the pump.

    (vi) Refer to Table 4-15, for tentatively sizing return piping.

    (vii) Tabulate the lengths of each section, calculate the heat loss as for the supply piping, and check against the assumptions and adjust if necessary.

    (viii) Convert the measured lengths of the return piping to equivalent lengths by multiplying by 1 1/2. Then calculate the friction losses from table 4-15. Total the friction losses from the longest (governing) run. Adjust pipe sizes if the friction loss in any section seems abnormally high (to save head on the pump). Sizes on nongoverning sections of the return can be reduced to save money where a pipe size smaller does not produce excessive friction.

(ix) Select the pump (circulator) to give at least the required gpm capacity with at least the required head and having a completely non-overloading curve.

(x) See Calculation Form J, Chapter 5.

Table 4-14 Heat loss Btu/(h)(lin ft)
(1-in thick insulation, 70°F ambient)

Pipe size	water temperature	
	140°F	190°F
1/2	6.9	12.3
3/4	8.2	14.7
1	8.5	15.3
1-1/4	11.2	20.1
1-1/2	11.4	20.4
2	12.9	23.2
2-1/2	14.8	26.7
3	17.6	31.6
4	21.1	37.9
5	25.7	46.2
6	30.6	55.0
8	36.8	66.1
10	47.0	84.6

Table 4-15 Friction loss in feet per 100 feet[a]

Size, in	gpm																							
	1/2	1	1-1/2	2	3	4	5	6	7	8	9	10	12	15	20	25	30	35	40	45	50	60	70	80
1/2	7	2.5*	5.3	8.9																				
3/4					1.5	3.2*	5.4	8.2																
1							1.5	2.2	3.1*	4.2	5.4	6.6	8.0											
1-1/4									1.5	1.9	2.4	2.9*	4.0											
1-1/2												1.3	1.7	2.6*	4.5	6.7								
2															1.2	1.8	2.5*	3.3	4.2	5.2	6.3	8.9		
2-1/2																		1.1	1.5	1.8	2.2	3.0*	4.1	5.3

*For initial sizing of possible governing runs use gpm figures with asterisks. To reduce abnormally high friction on sections of the governing run, use gpm to the left of figures with asterisks. For economy on nongoverning runs, use gpm to the right of figures with asterisks.

Table 4-16 Maximum allowable lengths of dead-leg hot-water branches

1. Maximum allowable lengths of dead-leg hot-water branches:

    a. 2-gpm (regular lavatory faucets and sinks:

        (1) One or two fixtures (3/4 in) should be 50 ft.
        (2) Batteries of fixtures (1 in and larger) should be 35 ft.

    b. 1/2-gpm lavatory faucets:

        (1) One or two fixtures (3/8 in) should be 40 ft.
        (2) Batteries of fixtures (1/2 in and larger) should be 25 ft.

2. For ceiling branches feeding down, 9 ft of allowable length must be assumed for the drop.

3. For ceiling branches feeding up, 3 ft of allowable length must be assumed for the rise.

4. Maximum allowable length should include the entire length (including ups and downs) from the tee in the circulating pipe to the tee feeding out through the wall at the farthest fixture, including both ceiling piping and piping in wall behind fixtures.

Table 4-17 Hot-water demand in fixture units (140°F water)

	Apartment house	Club	Gymnasium	Hospital	Hotels and dormitories	Industrial plant	Office building	School	YMCA
Basins, private lavatory	3/4	3/4	3/4	3/4	3/4	3/4	3/4	3/4	3/4
Basins, public lavatory		1	1	1	1	1	1	1	1
Bathtubs	1-1/2	1-1/2		1-1/2	1-1/2				
Dishwashers	1-1/2			\multicolumn{6}{l}{Five fixture units per 250 seating capacity}		
Therapeutic bath				5					
Kitchen sink	3/4	1-1/2		3	1-1/2	3		3/4	3
Pantry sink		2-1/2		2-1/2	2-1/2			2-1/2	2-1/2
Slop sink	1-1/2	2-1/2		2-1/2	2-1/2	2-1/2	2-1/2	2-1/2	2-1/2
Showers*	1-1/2	1-1/2	1-1/2	1-1/2	1-1/2	3		1-1/2	1-1/2
Circular wash fountain		2-1/2	2-1/2	2-1/2		4		2-1/2	2-1/2
Semicircular fountain		1-1/2	1-1/2	1-1/2		3		1-1/2	1-1/2

*In applications where principal use is showers, as in gymnasiums or at end of shift in industrial plants, use conversion factor of 1.00 to obtain design water flow rate in gpm.

Source: Reprinted, by permission, from ASHRAE Guide and Data Book, 1970.

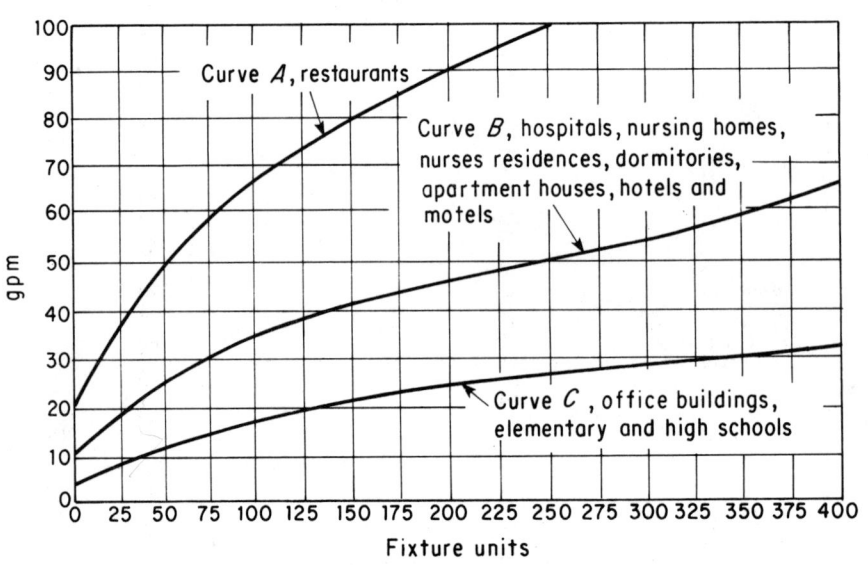

Fig. 4-7. Modified Hunter curve for hot-water flow rate (up to 400 fixture units).

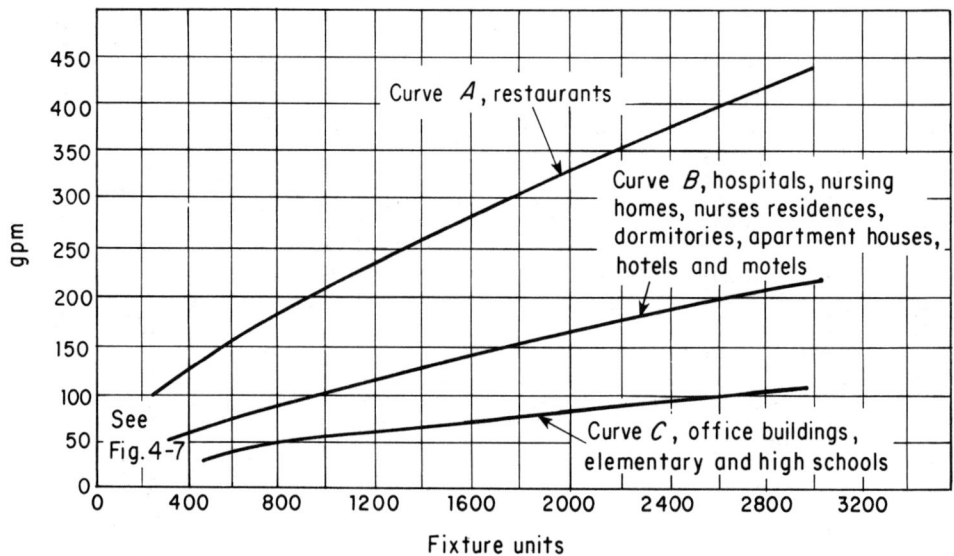

Fig. 4-8. Modified Hunter curve for hot-water flow rate (up to 3,000 fixture units).

Table 4-18 Hot-water demands and use for various types of buildings

Type of building	Maximum hour	Maximum day	Average day
Men's dormitories	3.8 gal/student	22.0 gal/student	13.1 gal/student
Women's dormitories	5.0 gal/student	26.5 gal/student	12.3 gal/student
Motels: no. of units[a]			
20 or less	6.0 gal/unit	35.0 gal/unit	20.0 gal/unit
60	5.0 gal/unit	25.0 gal/unit	14.0 gal/unit
100 or more	4.0 gal/unit	15.0 gal/unit	10.0 gal/unit
Nursing homes	4.5 gal/bed	30.0 gal/bed	18.4 gal/bed
Office buildings	0.4 gal/person	2.0 gal/person	1.0 gal/person
Food-service establishments:			
Type A—Full-meal restaurants and cafeterias	1.5 gal/(max meals)/(hr)	11.0 gal/(max meals)/(day)	2.4 gal/(avg meals)/(day)[b]
Type B—Drive-ins, grilles, luncheonettes, sandwich, and snack shops	0.7 gal/(max meals)/(hr)	6.0 gal/(max meals)/(day)	0.7 gal/(avg meals)/(day)[b]
Apartment houses: no. of apartments			
20 or less	12.0 gal/apt	80.0 gal/apt	42.0 gal/apt
50	10.0 gal/apt	73.0 gal/apt	40.0 gal/apt
75	8.5 gal/apt	66.0 gal/apt	38.0 gal/apt
100	7.0 gal/apt	60.0 gal/apt	37.0 gal/apt
130 or more	5.0 gal/apt	50.0 gal/apt	35.0 gal/apt
Elementary schools	0.6 gal/student	1.5 gal/student	0.6 gal/student[b]
Junior and senior high schools	1.0 gal/student	3.6 gal/student	1.8 gal/student[b]

[a] Interpolate for intermediate values
[b] Per day of operation

Above information reprinted, by permission, from ASHRAE Guide & Data Book, 1970.

Table 4-19 Hot-water demand per fixtures for various types of buildings
(Gallons of water per hour per fixture, calculated at a final temperature of 140°F)

	Apartment house	Club	Gymnasium	Hospital	Hotel	Industrial plant	Office Bldg	Private resident	School	YMCA
Basins, private lavatory	2	2	2	2	2	2	2	2	2	2
Basins, public lavatory	4	6	8	6	8	12	6	—	15	8
Bathtubs	20	20	30	20	20	—	—	20	—	30
Dishwashers[b]	15	50–150	—	50–150	50–200	20–100	—	15	20–100	20–100
Foot basins	3	3	12	3	3	12	—	3	3	12
Kitchen sink	10	20	—	20	30	20	20	10	20	20
Laundry, stationary tubs	20	28	—	28	28	—	—	20	—	28
Pantry sink	5	10	—	10	10	—	10	5	10	10
Showers	30	150	225	75	75	225	30	30	225	225
Stop sink	20	20	—	20	30	20	20	15	20	20
Hydrotherapeutic showers	—	—	—	400						
Hubbard baths	—	—	—	600						
Leg baths	—	—	—	100						
Arm baths	—	—	—	35						
Sitz baths	—	—	—	30						
Continuous-flow baths	—	—	—	165						
Circular wash sinks	—	—	—	20	20	30	20	—	30	
Semicircular wash sinks	—	—	—	10	10	15	10	—	15	
Demand factor	0.30	0.30	0.40	0.25	0.25	0.40	0.30	0.30	0.40	0.40
Storage capacity factor[c]	1.25	0.90	1.00	0.60	0.80	1.00	2.00	0.70	1.00	1.00

[a] Above information reprinted by permission from ASHRAE Handbook & Product Directory, 1973.
[b] Dishwasher requirements should be taken from manufacturer's data for the model to be used, if this is known.
[c] Ratio of storage-tank capacity to probable maximum demand per hour. Storage capacity may be reduced where an unlimited supply of steam is available from a central street steam system or large boiler plant.

Table 4-20 Plumbing fixture hot-water ratings

| | | Fixture Units | | | | gph | |
| | | | | (P-K) | ASHRAE | P-K | |
Fixture type	Symbol	FU	(Aerco)	400	(modified)	Storage	500
		Office buildings					
Bathroom group private	BGH	2.25	2.25	2.2	32	14	16
Public lavatory	LVH	1.5	1.0	1.0	6	2	2
Sink, light duty private	SPH	1.5	.75	1.5	2	2	2
Slop sink	SSH	2.25	2.5	2.0	20	15	10
Mop sink	MSH	2.25	2.5	2.0	20	15	10
Sink, light duty community pantry	SH	1.5	2.5	2.5	10	10	10
Public shower	SHH	3.0	2.5	2.5	30	15	15
Kitchen sink, heavy duty	KSH	3.0	3.0	3.0	20	15	12
Private shower	PSH	1.5	1.5	1.5	30	12	15
Private lavatory	PLH	0.75	0.75	0.7	2	2	1
		Industrial plants					
Public lavatory	LVH	1.5	1.0	2.0	12	10	10
Sink, light duty private	SPH	1.5	1.5	1.5	2	2	2
Slop sink	SSH	2.25	2.5	2.0	20	10	10
Mop sink	MSH	2.25	2.5	2.0	20	10	10
Sink, light duty community pantry	SH	1.5	2.5	2.5	10	10	10
Public shower	SHH	3.0	3.0	3.0	225	75	75
Kitchen sink, heavy duty	KSH	3.0	3.0	3.0	13	20	12
Private shower	PSH	1.5	1.5	1.5	30	12	12
Private lavatory	PLH	0.75	0.75	0.7	2	2	2
Laboratory sink	LSH	1.5	0.75	1.5	6	5	5
Cup drain	CDH	0.5	0.75	1.5	6	5	5
Laboratory outlet	LOH	0.5	0.75	1.5	6	5	5
Circular wash sink 36 in	CWH	...	4.0	1.5	30	240	100
Semicircular wash 36 in	SWH	...	3.0	1.0	15	200	60

(Continued on next page)

Table 4-20 (Continued)

		Fixture Units				gph	
Fixture type	Symbol	FU	(Aerco)	(P-K) 400	ASHRAE (modified)	P-K Storage	500

Hospitals and laboratories

Fixture type	Symbol	FU	(Aerco)	(P-K) 400	ASHRAE (modified)	P-K Storage	500
Bathroom group private	BGH	2.25	2.25	2.5	22	14	11
Flushing rim slop sink	FSH	2.25	2.5	2.0	20	10	8
Water closet with bedpan washer	WBH	2.25	2.5	2.0	10	10	8
Public lavatory	LVH	1.5	1.0	1.5	6	2	2
Sink, light duty private	SPH	1.5	1.0	1.5	2	2	2
Slop sink	SSH	2.25	2.5	2.5	20	10	10
Mop sink	MSH	2.25	2.5	2.5	20	10	10
Sink, light duty community pantry	SH	1.5	2.5	2.5	10	10	10
Public shower	SHH	3.0	1.5	2.5	75	20	15
Public bath with shower	BSH	3.0	1.5	2.5	75	15	15
Public bath, no shower	BH	3.0	1.5	2.0	20	20	20
Kitchen sink, heavy duty	KSH	3.0	3.0	3.0	20	15	12
Private shower	PSH	1.5	1.5	1.5	75	12	10
Private bath	PBH	1.5	1.5	1.5	20	12	10
Private lavatory	PLH	0.75	0.75	1.0	2	2	1
Laboratory sink	LSH	1.5	0.75	1.5	6	5	5
Cup drain	CDH	0.5	0.75	1.5	6	5	5
Laboratory outlet	LOH	0.5	0.75	1.5	6	5	5
Sill cock	SCH	2.25	2.5	2.5	20	10	10
Private bathroom group flush tank	PBH	2.25	2.5	2.2	22	14	11
Circular wash sink 36 in	CWH	4.0	4.0	1.5	20	240	100
Semicircular wash 36 in	SWH	3.0	3.0	1.0	10	200	60
Autopsy table	ATH	2.0	2.5	2.5	20	20	20
Autopsy sink	ASH	2.5	2.0	2.0	20	20	20
Arm bath	ARH	3.0	3.0	4.0	35	35	25
Emergency bath	EMH	3.0	3.0	2.0	165	165	20
Hubbard bath	HUH	15.0	10.0	10.0	600	600	400
Leg bath	LGH	7.0	4.0	4.0	100	100	80
Sitz bath	SIH	3.0	3.0	3.0	30	30	25
Foot basin	FTH	2.0	2.0	3.0	15	15	15
Bidet	BIH	3.0	1.0	1.0	2	2	2
Sonic cleaner	SOH	2.5	2.5	2.5	20	20	20
Laundry tub	LTH	2.5	2.5	1.5	35	35	15
Barber lavatory	BLH	1.5	1.5	2.0	15	15	12
Scrubup lavatory	SLH	1.5	1.5	1.5	10	10	6
Treatment lavatory	TRH	0.75	0.75	0.75	4	4	4
Shower obstetric	OSH	3.0	2.5	2.5	50	50	30
Shower theraputic	THH	11.0	5.0	8.0	400	400	220
Sanitizer, boiling type	SAN	2.0	2.0	2.0	10	10	7
Sink, barium	BSH	2.5	1.0	1.0	15	15	10
Sink, cleanup RM	CUH	2.5	1.0	1.0	15	15	10
Sink, plaster	PLH	2.5	1.0	1.0	15	15	10
Sink, central supply	CSH	2.5	1.0	1.0	15	15	10

(Continued on next page)

Table 4-20 (Continued)

		Fixture Units				gph	
				(P-K)	ASHRAE (modified)	P-K	
Fixture type	Symbol	FU	(Aerco)	400		Storage	500

Hospitals and laboratories (Contd)

Fixture type	Symbol	FU	(Aerco)	(P-K) 400	ASHRAE (modified)	Storage	500
Sink, substerilizer	SRH	2.5	1.0	1.0	15	15	10
Sink, clean utility	CLH	2.5	1.0	1.0	15	15	10
Sink, kitchen floor	FKH	3.0	2.5	2.5	20	20	20
Sink, formula room	FOH	3.0	2.0	2.0	20	20	20
Sink, pharmacy and nurses	PHH	1.5	1.5	1.5	5	5	2
Scrubup sink	SUH	3.0	1.5	1.5	50	50	30
Soiled utility sink	UTH	2.5	1.0	1.0	20	20	15
Water sterilizer	STH	2.0	2.0	2.0	10	10	10
Flask washer	FLK	4.0	1.5	1.5	10	10	10
Glove washer	GLH	3.0	1.5	1.5	20	20	20
Pipette washer	PPH	3.0	1.5	1.5	15	15	15
Utensil sterilizer	UTH	1.5	1.5	1.5	10	10	10
Prerinse hose	PRH	2.5	2.5	2.5	45	45	45
Bar sink, soak sink	BAH	2.5	2.5	2.5	30	30	30
Pot sink/faucet	PSH	3.0	2.5	2.5	30	30	30
Vegetable sink	VEH	2.5	2.0	2.0	45	45	45
Soup kettle	SKH	2.5	2.5	2.5	60	60	45
Can washer	CNH	10.0	10.0	10.0	50	50	50
Kitchen glassware washer	GLH	1.5	1.5	2.0	45	45	35
Hospital glass washer	GLW	6.0	6.0	6.0	180	180	135
Semiprivate bath	SBH	2.0	1.5	1.5	30	30	12
Semiprivate shower	SEH	2.0	1.5	1.5	20	20	12
Semiprivate lavatory	SLH	1.5	1.2	1.2	4	4	1
Pot and pan washer	POT	3.0	2.0	2.0	100	100	80
Silver washer	SIL	2.0	2.0	2.0	45	45	40
Syringe washer	SYR	4.0	2.0	2.0	45	45	40

Medical buildings

Fixture type	Symbol	FU	(Aerco)	(P-K) 400	ASHRAE (modified)	Storage	500
Public lavatory	LVH	1.5	1.0	1.0	6	3	2
Sink, light duty private	SPH	1.5	1.0	1.5	2	2	2
Slop sink	SSH	2.25	2.5	2.0	20	10	10
Mop sink	MSH	2.25	2.5	2.0	20	10	10
Flushing rim slop sink	FSH	2.25	2.5	2.0	20	10	8
Private shower	PSH	1.5	1.5	1.5	75	20	12
Private lavatory	PLH	0.75	0.75	0.7	2	2	1
Laboratory sink	LSH	1.5	0.75	1.5	6	5	5
Cup drain	CDH	0.5	0.75	1.5	6	5	5
Laboratory outlet	LOH	0.5	0.75	1.5	6	5	5
Arm bath	ARH	3.0	3.0	4.0	35	35	25
Leg bath	LGH	7.0	4.0	4.0	100	100	80
Sitz bath	SIH	3.0	3.0	3.0	30	30	25
Foot basin	FTH	2.0	2.0	3.0	15	15	15
Bidet	BIH	3.0	1.0	1.0	2	2	2

(Continued on next page)

Table 4-20 (Continued)

		Fixture Units				gph	
Fixture type	Symbol	FU	(Aerco)	(P-K) 400	ASHRAE (modi-fied)	P-K Storage	500

Correctional buildings

Fixture type	Symbol	FU	(Aerco)	(P-K) 400	ASHRAE (mod)	P-K Storage	500
Bathroom group private	BGH	2.25	2.25	2.2	22	17	21
Flushing rim slop sink	FSH	2.25	2.5	2.0	20	15	8
Water closet with bedpan washer	WBH	2.25	2.5	2.0	10	10	8
Public lavatory	LVH	1.5	1.0	1.0	6	2	2
Sink, light duty private	SPH	1.5	1.0	1.5	2	2	2
Slop sink	SSH	2.25	2.5	2.0	20	15	10
Mop sink	MSH	2.25	2.5	2.0	20	15	10
Sink, light duty community pantry	SH	1.5	2.5	2.5	10	15	10
Public shower	SHH	3.0	1.5	3.0	75	15	75
Public bath with shower	BSH	3.0	1.5	3.0	75	20	75
Public bath, no shower	BH	3.0	1.5	2.0	20	15	15
Kitchen sink, heavy duty	KSH	3.0	3.0	3.0	20	20	15
Private shower	PSH	1.5	1.5	1.5	75	12	12
Private bath	PBH	1.5	1.5	1.5	20	12	20
Private lavatory	PLH	0.75	0.75	0.7	2	2	1

Private residences

Fixture type	Symbol	FU	(Aerco)	(P-K) 400	ASHRAE (mod)	P-K Storage	500
Bathroom group private	BGH	2.25	none listed	none listed	22	14	none listed
Sink, light duty private	SPH	1.5	...	...	5	5	...
Slop sink	SSH	2.25	...	...	15	2	...
Mop sink	MSH	2.25	...	...	15	2	...
Sink, light duty community pantry	SH	1.5	...	...	5	5	...
Kitchen sink, heavy duty	KSH	3.0	...	...	10	10	...
Private shower	PSH	1.5	...	...	30	12	...
Private bath	PBH	1.5	...	...	20	12	...
Private lavatory	PLH	0.75	...	...	...	2	...
Private bathroom, group flush tank	PBH	2.25	...	...	22	14	...
Bidet	BIH	3.0	1.0	1.0	2	2	...

(Continued on next page)

Table 4-20 (Continued)

		Fixture Units				gph	
Fixture type	Symbol	FU	(Aerco)	(P-K) 400	ASHRAE (modified)	P-K Storage	500
Apartment buildings							
Bathroom group private	BGH	2.25	2.25	2.2	22	14	14
Public lavatory	LVH	1.5	1.0	1.0	4	2	2
Sink, light duty private	SPH	1.5	.75	1.5	2	5	5
Slop sink	SSH	2.25	1.5	1.5	20	8	8
Mop sink	MSH	2.25	1.5	1.5	20	8	8
Sink, light duty community pantry	SH	1.5	0.75	1.5	5	5	5
Kitchen sink, heavy duty	KSH	3.0	0.75	1.5	10	8	5
Private shower	PSH	1.5	1.5	1.5	30	12	12
Private bath	PBH	1.5	1.5	1.5	20	12	12
Private lavatory	PLH	0.75	0.75	0.7	2	2	2
Private bathroom group flush tank	PBH	2.25	2.25	2.2	22	14	17
Bidet	BIH	3.0	1.0	1.0	2	2	2
Hotels and dormitories							
Bathroom group private	BGH	2.25	2.25	2.2	77	14	14
Public lavatory	LVH	1.5	1.0	1.0	8	3	3
Sink, light duty private	SPH	1.5	1.5	1.5	2	2	2
Slop sink	SSH	2.25	2.5	2.5	30	10	10
Mop sink	MSH	2.25	2.5	2.5	30	10	10
Sink, light duty community pantry	SH	1.5	1.5	2.5	10	10	10
Public shower	SHH	3.0	2.5	2.5	75	20	12
Public bath with shower	BSH	3.0	1.5	1.6	75	12	12
Public bath, no shower	BH	3.0	1.5	2.0	20	12	12
Kitchen sink, heavy duty	KSH	3.0	2.5	3.0	30	15	12
Private shower	PSH	1.5	1.5	1.5	75	12	12
Private bath	PBH	1.5	1.5	1.5	20	12	12
Private lavatory	PLH	0.75	0.75	0.70	2	2	2
Private bathroom group flush tank	PBH	2.25	2.25	2.2	77	14	14

(Continued on next page)

Table 4-20 (Continued)

Fixture type	Symbol	Fixture Units			ASHRAE (modified)	gph	
		FU	(Aerco)	(P-K) 400		P-K Storage	500

Public schools

Fixture type	Symbol	FU	(Aerco)	(P-K) 400	ASHRAE (modified)	P-K Storage	500
Public lavatory	LVH	1.5	1.0	1.0	15	2	2
Sink, light duty private	SPH	1.5	1.5	1.5	2	2	2
Slop sink	SSH	2.25	2.5	2.0	20	8	10
Mop sink	MSH	2.25	2.5	2.0	20	8	10
Sink, light duty community pantry	SH	1.5	.75	2.5	20	20	10
Public shower	SHH	3.0	1.5	1.7	225	15	15
Kitchen sink, heavy duty	KSH	3.0	2.5	3.0	20	20	12
Private shower	PSH	1.5	1.5	1.7	15	15	15
Private lavatory	PLH	0.75	0.75	1.0	2	2	1
Laboratory sink	LSH	1.5	0.75	1.5	6	5	5
Cup drain	CDH	0.5	0.75	1.5	6	5	5
Laboratory outlet	LOH	0.5	0.75	1.5	6	5	5
Circular wash sink	CWH	4.0	4.0	1.5	30	240	100
Semicircular wash	SWH	3.0	3.0	1.0	15	200	60

Private schools

Fixture type	Symbol	FU	(Aerco)	(P-K) 400	ASHRAE (modified)	P-K Storage	500
Public lavatory	LVH	1.5	1.0	1.0	15	2	2
Sink, light duty private	SPH	1.5	1.5	1.5	2	2	2
Slop sink	SSH	2.25	2.5	2.0	20	8	10
Mop sink	MSH	2.25	2.5	2.0	20	8	10
Sink, light duty community pantry	SH	1.5	0.75	2.5	10	15	10
Public shower	SHH	3.0	1.5	1.7	225	15	15
Kitchen sink, heavy duty	KSH	3.0	2.5	3.0	20	15	15
Private shower	PSH	1.5	1.5	1.7	15	15	15
Private lavatory	PLH	0.75	0.75	1.0	2	2	1
Laboratory sink	LSH	1.5	0.75	1.5	6	5	5
Cup drain	CDH	0.5	0.75	1.5	6	5	5
Laboratory outlet	LOH	0.5	0.75	1.5	6	5	5
Circular wash sink	CWH	4.0	4.0	1.5	30	240	100
Semicircular wash	SWH	3.0	3.0	1.0	15	200	60

Gymnasiums

Fixture type	Symbol	FU	(Aerco)	(P-K) 400	ASHRAE (modified)	P-K Storage	500
Public lavatory	LVH	1.5	1.0	1.5	8	4	4
Sink, light duty private	SPH	1.5	1.5	1.5	2	2	2
Slop sink	SSH	2.25	2.5	1.5	20	10	8
Mop sink	MSH	2.25	2.5	1.5	20	10	8
Kitchen sink, heavy duty	KSH	3.0	3.0	3.0	20	15	12
Private shower	PSH	1.5	1.5	1.5	30	12	12

(Continued on next page)

Table 4-20 (Continued)

		Fixture Units				gph	
				(P-K)	ASHRAE (modified)	P-K	
Fixture type	Symbol	FU	(Aerco)	400		Storage	500

Gymnasiums (Contd)

Fixture type	Symbol	FU	(Aerco)	(P-K) 400	ASHRAE (modified)	Storage	500
Private lavatory	PLH	0.75	0.75	1.0	2	2	2
Circular wash sink	CWH	4.0	4.0	1.5	20	240	100
Semicircular wash	SWH	3.0	3.0	1.0	10	200	60
Arm bath	ARH	3.0	3.0	4.0	35	35	25
Leg bath	LGH	7.0	4.0	4.0	100	100	80

Clubs

Fixture type	Symbol	FU	(Aerco)	(P-K) 400	ASHRAE (modified)	Storage	500
Bathroom group private	BGH	2.25	2.25	2.3	22	14	13
Public lavatory	LVH	1.5	1.0	1.3	6	4	2
Sink, light duty private	SPH	1.5	0.75	1.5	2	2	2
Slop sink	SSH	2.25	2.5	2.5	20	10	8
Mop sink	MSH	2.25	2.5	2.5	20	10	8
Sink, light duty community pantry	SH	1.5	1.5	2.5	10	10	10
Public shower	SHH	3.0	1.5	1.7	150	75	15
Public bath with shower	BSH	3.0	1.5	1.6	20	75	15
Public bath, no shower	BH	3.0	1.5	1.7	20	12	12
Kitchen sink, heavy duty	KSH	3.0	2.5	3.0	20	20	12
Private shower	PSH	1.5	1.5	1.5	20	12	12
Private bath	PBH	1.5	1.5	1.5	20	12	12
Private lavatory	PLH	0.75	0.75	0.80	2	2	1
Sill cock	SCH	2.25	2.5	2.5	20	10	10
Private bathroom, group flush tank	PBH	2.25	2.25	2.3	22	14	13
Circular wash sink	CWH	4.0	4.0	1.5	20	240	100
Semicircular wash	SWH	3.0	3.0	1.0	10	200	60
Arm bath	ARH	3.0	3.0	4.0	35	35	25
Leg bath	LGH	7.0	4.0	4.0	100	100	80
Bidet	BIH	3.0	1.0	1.0	2	2	2

YMCA'S and YWCA'S

Fixture type	Symbol	FU	(Aerco)	(P-K) 400	ASHRAE (modified)	Storage	500
Bathroom group private	BGH	2.25	2.25	2.5	32	17	22
Public lavatory	LVH	1.5	1.0	1.0	8	3	3
Sink, light duty private	SPH	1.5	1.5	1.5	2	2	2
Slop sink	SSH	2.25	2.5	2.5	20	8	10
Mop sink	MSH	2.25	2.5	2.5	20	8	10
Sink, light duty community pantry	SH	1.5	2.5	2.5	10	10	10
Public shower	SHH	3.0	1.5	1.7	225	15	15

(Continued on next page)

Table 4-20 (Continued)

		Fixture Units				gph	
				(P-K)	ASHRAE (modified)	P-K	
Fixture type	Symbol	FU	(Aerco)	400		Storage	500

YMCA'S and YWCA'S (Contd)

Fixture type	Symbol	FU	(Aerco)	(P-K) 400	ASHRAE (modified)	Storage	500
Public bath with shower	BSH	3.0	1.5	1.7	225	20	20
Public bath, no shower	BH	3.0	1.5	1.7	30	15	20
Kitchen sink, heavy duty	KSH	3.0	3.0	3.0	20	15	15
Private shower	PSH	1.5	1.5	1.5	30	15	15
Private bath	PBH	1.5	1.5	1.7	30	15	20
Private lavatory	PLH	0.75	0.75	1.0	2	2	2
Sill cock	SCH	2.25	0.75	2.5	20	10	10
Private bathroom group flush tank	PBH	2.25	2.25	2.5	32	17	22
Circular wash sink	CWH	4.0	4.0	1.5	30	240	100
Semicircular wash	SWH	3.0	3.0	1.0	20	200	60

How to use these tables: In calculating service hot-water loads to size hot-water heaters, the characteristics of the hot-water heater has to be known. The three columns headed Fixture units provide FU data for semi-instantaneous or instantaneous heaters, while the three columns headed gph provide gallons-per-hour data for storage-type heaters. With this array of data, the plumbing engineer can select the system that best fits his project requirements.

Reprinted, by permission, from Building Systems Design.

Table 4-21 Proposed hospital plumbing fixture ratings

	Fixture Units			gpm		gph
Fixture	Total, water	Cold water	Hot water	Cold water	Hot water	Hot water
General areas						
Aspirator, fluid suction	2	2	....	3	....	....
Aspirator, laboratory	2	2	....	3	....	....
Autopsy table, complete	4	3	2	8	4-1/2	20
Autopsy table, aspirator	2	2	....	3	....	....
Autopsy table, flushing hose	2	2	....	3	....	....
Autopsy table, flushing rim	3	3	....	4-1/2	....	....
Autopsy table, sink and faucet	3	2-1/2	2-1/2	4-1/2	4-1/2	20
Autopsy table, waste disposal	1-1/2	1-1/2	....	4	....	....
Bath, arm	4	2	3	3	7	35
Bath, emergency	4	2	3	3	7	15
Bath, immersion	20	7	15	15	35	450
Bath, leg	10	4	7	8	16	100
Bath, sitz	4	2	3	3	7	30
Bathtub, general	4	2	3	3	7	45
Bathtub, private	2	1-1/2	1-1/2	3	7	15
Bedpan, washer, steam	10	10	....	25	....	....
Bidet	4	3	3	4-1/2	4-1/2	15
Cleaner, sonic	3	2-1/2	2-1/2	4-1/2	4-1/2	20
Cuspidor, dental and surgical	1	1	....	2	....	....
Cuspidor, dental chair	1	1	....	2	....	....
Drinking fountain	1	1	....	2	....	....
Floor drain, flushing type	10	10	....	25	....	....
Hose, bedpan general	2	1-1/2	1-1/2	3	3	5
Hose, bedpan private	1	1	1	3	3	2
Laundry tub	3	2-1/2	2-1/2	4-1/2	4-1/2	30
Lavatory, barber	2	1-1/2	1-1/2	3	3	15
Lavatory, dental	1	1	1	3	3	8
Lavatory, general	2	1-1/2	1-1/2	3	3	8
Lavatory, private	1	1	1	3	3	4
Lavatory, nursery	2	1-1/2	1-1/2	3	3	8
Lavatory, scrub-up	2	1-1/2	1-1/2	3	3	10
Lavatory, treatment	1	1	1	3	3	4
Microscope, electron	1	1	....	0.2	....	....
Sanitizer, boiling, instrument	2	....	2	....	3	10
Sanitizer, boiling, utensil	2	....	2	....	3	10

(Continued on next page)

Table 4-21 (Continued)

	Fixture Units			gpm		gph
Fixture	Total, water	Cold water	Hot water	Cold water	Hot water	Hot water

General areas (Contd)

Fixture	Total, water	Cold water	Hot water	Cold water	Hot water	Hot water
Shower, general	4	2	3	1-1/2	3-1/2	50
Shower, private	2	1	2	1-1/2	3-1/2	20
Shower, obstetrical	4	2	3	1-1/2	3-1/2	50
Shower, therapeutic	15	6	11	15	35	400
Sink, barium	3	2-1/2	2-1/2	4-1/2	4-1/2	15
Sink, clean-up room	3	2-1/2	2-1/2	4-1/2	4-1/2	15
Sink, central supply	3	2-1/2	2-1/2	4-1/2	4-1/2	15
Sink, clinical	10	10	3	25	3	10
Sink, clinical, bedpan hose	10	10	4	25	4-1/2	15
Sink, floor kitchen	4	3	3	4-1/2	4-1/2	20
Sink, formula room	4	3	3	4-1/2	4-1/2	20
Sink, cup	1	1	....	3	....	....
Sink, laboratory	2	1-1/2	1-1/2	3	3	5
Sink, laboratory and trough	3	2-1/2	1-1/2	5	3	5
Sink, pharmacy	2	1-1/2	1-1/2	3	3	5
Sink, plaster	4	3	3	4-1/2	4-1/2	15
Sink, private kitchen	3	2-1/2	2-1/2	4-1/2	4-1/2	15
Sink, miscellaneous general purpose	3	2-1/2	2-1/2	4-1/2	4-1/2	10
Sink, nurses station	2	1-1/2	1-1/2	3	3	5
Sink, scrubup	4	3	3	4-1/2	4-1/2	50
Sink, substerile room	3	2-1/2	2-1/2	4-1/2	4-1/2	15
Sink, clean utility	3	2-1/2	2-1/2	4-1/2	4-1/2	15
Sink, soiled utility	3	2-1/2	2-1/2	4-1/2	4-1/2	20
Sterilizer, autoclave	2-6	2-6	....	3-10	....	....
Sterilizer, boiling, instrument	2	....	2	....	3	10
Sterilizer, boiling, utensil	2	....	2	....	3	10
Sterilizer, pressure instrument	2	2	....	3	....	....
Washer-sterilizer	6	6	....	10	....	....
Sterilizer, water	5	5	2	3	4-1/2	15
Urinal, pedestal, flush valve	10	10	....	15	....	....
Urinal, stall, flush valve	5	5	....	6	....	....
Urinal, wall, flush valve	5	5	....	15	....	....
Urinal, tank	3	3	....	1	....	....
Washer, flask	4	....	4	....	4-1/2	10
Washer, formula bottle	4	4	....	5	....	....
Washer, glove	4	3	3	4-1/2	4-1/2	20
Washer, glassware	4-6	....	4-6	....	5-10	25-180
Washer, needle	2	2	....	3	....	....
Washer, pipette	4	3	3	4-1/2	4-1/2	15
Washer, syringe	4	....	4	....	5	10

(Continued on next page)

Table 4-21 (Continued)

Fixture	Fixture Units			gpm		gph
	Total, water	Cold water	Hot water	Cold water	Hot water	Hot water
General areas (Contd)						
Washer, tube	4	4	....	5	....	....
Washer, sterilizer, utensil	2	1-1/2	1-1/2	3	3	10
Water closet, flush valve, general	10	10	....	25	....	....
Water closet, flush valve, private	6	6	....	25	....	....
Water closet, tank, general	5	5	....	3	....	....
Water closet, tank, private	3	3	....	3	....	....
Kitchen area						
Baine Marie	2	2	....	3	....	....
Coffee urn	2	2	....	3	....	....
Grinder, food waste	3	3	....	8	....	....
Hose, prerinse	3	2-1/2	2-1/2	4	4	45
Lavatory	2	1-1/2	1-1/2	3	3	10
Peeler, vegetable	3	3	....	4-1/2	....	....
Sink, bar type	3	2-1/2	2-1/2	4	4	30
Sink, dish soak	3	2-1/2	2-1/2	4	4	30
Sink, pot and pan (per faucet)	4	3	3	4-1/2	4-1/2	30
Sink, vegetable	3	2-1/2	2-1/2	4	4	45
Soup, kettle	2	1-1/2	1-1/2	3	3	60
Steam table	2	2	....	3	....	....
Washer, can	10	10	10	15	15	50
Washer, glassware	2	1-1/2	1-1/2	3	3	15
Washer, pot and pan	3	....	3	....	10	75
Washer, silver	2	....	2	....	4-1/2	45
Special areas						
Condenser, drinking fountain	1	1		1		
Condenser, refrigeration	1	1		1		
Condenser, sterilizer	2-6	2-6		3-10		
Cooling coil, water sterilizer	5	2		3		
Cooling tower makeup	....	....		.1/Ton		
Fire sprinklers	10	10		20		
Ice cuber and flaker	1	1		1		
Hose connection, interior	4	4		5		
Hose connection, sill cock	4	4		5		
Hose connection and hot-water supply	4	3	3	4-1/2	4-1/2	15
Pump, air compressor	1-8	1-8		1/2-12		
Pump, vacuum	1-8	1-8		1/2-12		
Still	1-6	1-6		1/2-10		

Source: Reprinted, by permission, from Air Conditioning, Heating & Ventilating (now Building Systems Design).

Table 4-22 Residential sizing

Introduction

A. O. Smith residential water heaters are produced in a large variety of tank sizes and heat inputs to permit the selection of the one best suited to do the job. Ideally this heater would have a combination of storage and heat input equal to the usage. In addition to the design factors and the sizing examples which follow, an appendix section provides detailed explanations of selected terminology. This is done to avoid expanding the content of the sizing procedure.

Design factors

These design factors are the result of combining A. O. Smith engineering test data and practical experience to form a usable guide for the selection of minimum water-heater tank sizes and heat inputs. As stated previously, the factors may be adjusted to suit individual needs.

1. 2-hour peak-usage period. Residential peak usage, based on accepted practice, is the 2-hour period during the day when the heaviest draw of hot water will occur. For example, from 7:00 to 9:00 A.M.

2. Gallons of 140°F hot water required: 20 gal/person for the first two persons; 5 gal/person for each person over the first two; 10 gal for each full bath over the first bath; 10 gal for an automatic dishwasher; 20 gal for an automatic clothes washer.

3. Storage tank-size selection. (Note: The draw efficiency of a gas or electric water heater storage tank is considered to be 70 percent.) 30-gal size (21-gal draw) for one bath residence; 40-gal size (28-gal draw) for two-bath residence, or one bath with an automatic clothes washer; 50-gal size (35-gal draw) for three-bath residence, or two baths with an automatic clothes washer.

4. Oil-fired water heaters with their higher recovery normally can use a smaller storage tank than a gas or electric model.

5. Heat input VS recovery capacity. Oil and gas water-heater recovery table (calculated at 70 percent recovery efficiency).

Input rating Btu/hour	gph recovery at indicated temperature rise				
	60°	70°	80°	90°	100°
18,000	25.2	21.6	18.9	16.8	15.1
20,000	28.0	24.0	21.0	18.7	16.8
25,000	35.0	30.0	26.2	23.3	21.0
30,000	42.0	36.0	31.5	28.0	25.2
33,000	46.2	39.6	34.7	30.2	27.7
35,000	49.0	42.0	36.8	32.7	29.4
40,000	56.0	47.6	42.7	37.1	33.6
43,000	60.2	51.6	45.2	40.2	36.1
50,000	70.0	60.0	52.5	46.7	42.0
60,000	84.0	72.0	63.1	56.0	50.4
70,000	98.0	84.0	73.5	65.1	58.8
80,000	112.0	96.0	84.0	74.4	67.2
90,000	126.0	108.0	94.5	83.7	75.6
100,000	140.0	120.0	105.0	93.0	84.0

Source: Reprinted, by permission, A. O. Smith Corp., Consumer Products Division, Kankakee, Ill.

Table 4-23 Hot-water requirements – apartments

This table has been prepared to serve as a guide for estimating the 3-hour hot-water demand for various-sized apartment buildings. Minimum storage capacities are also shown. The table assumes an average occupancy of 2-1/2 persons per apartment and 5-minute showers. (Note: Estimated 3-hour demands shown include shower and other minor uses such as lavatories and residential dishwashers. Other major hot-water-consuming appliances such as clothes washers will increase the total demand. Consult manufacturers specifications for hot-water consumption and increase generating and storage capacity accordingly. Diversity factors have been used in calculating expected hot-water requirements. Calculations below are based on 3-gpm shower flow rate of mixed temperature water. If shower flow is up to 5 gpm, multiply gallon requirements in chart by 1.6.

(1)	(2)	(3)		(4)
		Gallons required 3-hour period, 140°F water		
Number of apartments (2-1/2 persons/apt.)	Actual number of persons	40°F inlet 100°TR	60°F inlet 80°TR	Minimum storage capacity[a]
1-3	7	85	72	50
4	10	110	94	60
5-6	15	164	140	72
7-8	20	220	185	85
9-10	25	275	232	100
11-15	37	375	319	113
16-20	50	500	425	130
21-25	62	620	527	148
26-30	75	750	638	162
31-35	87	870	740	175
36-40	100	1000	850	188
41-45	112	1042	886	200
46-50	125	1100	935	210
51-75	187	1552	1320	255
76-100	250	1975	1680	300
101-125	312	2370	2015	325
126-150	375	2775	2359	360
151-175	437	3146	2674	395
176-200	500	3500	2975	410
201-250	625	4125	3506	500
251-300	750	4500	3825	600
301-350	875	5250	4463	700

[a] Storage capacities shown are theoretical minimums.
For conditions other than those stated above consult your A. O. Smith supplier.

To use this table:

1. Determine number of apartments from column 1.
2. Determine number of occupants from column 1. (Note: If average occupancy differs from 2-1/2 persons per unit, disregard column 1 and use column 2 for estimating 3-hour demand and minimum storage capacity.)
3. Read expected 3-hour demand from column 3 for either 40° or 60°F inlet temperature.
4. Read minimum-system storage capacity from column 4.
5. Consult appropriate availability table for equipment selection. (Be sure storage capacity of system selected is no less than shown in column 4.)

Source: Reprinted, by permission, A. O. Smith Corp., Consumer Products Division, Kankakee, Ill.

Table 4-24 Hot-water requirements - motels and hotels

This table may be used as a guide for estimating the 2-hour hot-water demand for various-sized motels and hotels. Minimum storage capacities are also shown. The table assumes an average occupancy of 1-1/2 persons per unit and 5-minute showers. (Note: Hot-water load for restaurants, laundry operations, or other uses should be considered separately.) Diversity factors have been used in calculating expected hot water requirements. Calculations below based on 3-gpm shower flow rate of mixed temperature water. If shower flow is up to 5 gpm, multiply gallon requirements in chart by 1.6.

(1)	(2)	(3)		(4)
		Gallons required 2-hour period, 140°F water		
Number of units (1-1/2 persons/unit)	Actual number of persons	40°F inlet 100°TR	60°F inlet 80°TR	Minimum storage capacity[a]
1-3	4	50	45	50
4	6	66	56	60
5-6	9	100	85	72
7-8	12	132	112	85
9-10	15	165	140	100
11-15	22	230	196	113
16-20	30	300	255	130
21-25	37	370	315	148
26-30	45	450	382	162
31-35	52	520	442	175
36-40	60	570	485	188
41-45	67	600	510	200
46-50	75	650	552	210
51-75	112	840	714	255
76-100	150	1050	892	300
101-125	187	1272	1080	325
126-150	225	1350	1148	360
151-175	262	1575	1340	395
176-200	300	1800	1530	410
201-250	375	2250	1912	500
251-300	450	2700	2295	600
301-350	525	3150	2678	700

[a] Storage capacities shown are theoretical minimums. Motels and hotels with convention facilities and/or along busy interstate highways will commonly have a 1-hour peak period. Use the One-Hour Availability Tables for sizing.

For conditions other than those stated above consult your A. O. Smith supplier.

To use this table:

1. Determine number of units from column 1.
2. Determine number of persons from column 2. (Note: If average occupancy differs from 1-1/2 persons per unit, disregard column 1 and use column 2 for estimating 2-hour demand and minimum storage capacity.
3. Read estimated 2-hour demand from column 3 for either 40° or 60°F inlet temperature.
4. Read minimum storage capacity from column 4.
5. Consult appropriate availability table for equipment selection. (Be sure storage capacity of system selected is no less than shown in column 4.)

Source: Reprinted, by permission, A. O. Smith Corp., Consumer Products Division, Kankakee, Ill.

Table 4-25 Hot-water requirements - dormitories

This table may be used as a guide for estimating the 1-hour hot-water demand for either women's or men's dormitories. Estimated 1-hour usages are taken from the EEI studies which showed a peak hourly load of 3.8 gal of hot water per person for men's dormitories and 5.0 gal of hot water per person for women's dormitories. Shower head flow rates should be restricted to 3 or 4 gpm of mixed temperature water.

(1)	(2)				(3)
	Gallons required 1-hour period, 140°F water				
	Men		Women		
Number of persons	40°F inlet 100°TR	60°F inlet 80°TR	40°F inlet 100°TR	60°F inlet 80°TR	Minimum storage capacity[a]
1-10	95	76	125	100	100
11-15	141	113	187	150	150
16-20	190	152	250	200	200
21-25	210	168	277	220	225
26-30	222	178	300	240	240
31-40	253	203	320	264	280
41-50	266	223	350	280	310
51-75	313	252	412	330	400
76-100	380	304	500	400	430
101-125	475	380	625	500	475
126-150	570	456	750	600	510
151-175	665	532	875	700	560
176-200	760	610	1000	800	600
201-250	950	760	1250	1000	650
251-300	1140	913	1500	1200	720
301-350	1330	1065	1750	1400	800

[a] Storage capacities shown are theoretical minimums.

To use this table:

1. Determine number of persons from column 1.
2. Read gallons of 140°F water required from column 2 for either men's or women's dormitories at 40° or 60°F inlet temperature.
3. Read minimum storage capacity required from column 3.
4. Consult appropriate availability table for equipment selection. (Be sure storage capacity of system selected is no less than shown in column 3.)

Source: Reprinted, by permission, A. O. Smith Corp., Consumer Products Division, Kankakee, Ill.

5. Chilled-drinking-water systems.

   a. Principles of design.

   (1) The minimum number of drinking fountains (water coolers) required in a building or area of a building is usually governed by code. Most codes today also have requirements regarding drinking fountains for the handicapped.
   (2) Types of drinking fountains include:

   (a) Projecting type.
   (b) Semirecessed type.
   (c) Recessed type.
   (d) Simulated recessed type.
   (e) Unit water coolers.
   (f) Cafeteria water coolers.

   (3) Drinking fountains for the handicapped must be projecting type, with projection of at least 18 in and mounted 34 to 36 in above the floor.
   (4) Provide all drinking water outlets with chilled water at 50°F.
   (5) Where drinking fountains are stacked in the same location and the equipment space is reasonably adjacent, consider using central (or subcentral) circulating chilled-water systems as they will be economically feasible.
   (6) In areas where drinking fountain locations are scattered or irregular, provide local chilled-water units.
   (7) Build local chilled-water units either into the wall under the drinking fountain, exposed in a closet behind the drinking fountain, into the bottom of the semirecessed drinking fountain, into the back of the surface-mounted drinking fountain, or use unit water coolers, as per the architect's preference in each location. The units built into the wall or exposed in a closet may also feed drinking fountains on the floor above and/or below in the same location.
   (8) Whenever available, provide refrigerated drinking fountains (water coolers) with a precooler to save on energy costs of producing chilled water.

(9) Unit water coolers are sometimes provided with an integral hot-water unit and dispens-faucet to provide hot water for making instant coffee.
(10) Cafeteria water coolers are usually provided with push-back glass filling faucets. Sometimes they are also provided with a bubbler.
(11) All central and subcentral chilled-water systems should be circulating type with all bronze pumps.
(12) Where more than one riser is required, the circulation line can act as one of the feed risers.
(13) In up-feed risers, connection to the circulating line should be taken, where possible, at the ceiling below the top branch.

b. Equipment.

(1) All local chilled-water units should be air-cooled.
(2) The central chilled-water units should be air-cooled (whenever possible), air-cooled with blower; air- and water-cooled, or water-cooled, as size, location, and the requirements of the local authorities dictate.
(3) The central units should be sized to provide the required 45°F water to serve the people using the system, plus compensation for the circulating piping heat gain and the pump heat input.

   (a) Where the number of people involved cannot be determined, use 5 gph/drinking fountain.
   (b) The capacity required for people served: Permanent population = 1 gph/20 people; places of assembly or transient population = 5 gph/drinking fountain.
   (c) The heat gain in gph/100 ft of piping, assuming 45°F circulating water, 80°F inlet water, and a 90°F room, is: for 1/2 in, 2.0; for 3/4 in, 2.2; for 1 in, 2.3; and for 1 1/4 in, 2.4.
   (d) The heat gain from the circulating pump in gph, assuming 45°F circulating water, 80°F inlet water, and a 90°F room, is: for 1/4 hp, 2.2; for 1/3 hp, 2.9; for 1/2 hp, 4.4;

for 3/4 hp, 6.5; and for 1 hp, 8.7.

(e) Select the unit on the basis of the required gph (80°F to 45°F). The ratio of gph to storage is fixed by the manufacturer.

(f) Consult the manufacturer about the compressor for his unit, depending on the type of cooler selected.

(g) The circulating pump should be the gpm selected by the manufacturer for his unit, with the head a total of his required head plus the head of the reduced gpm circulated through the system.

(4) Base the required system circulation (gpm) on a minimum of 1 gpm or on the following gpm's/100 ft of circulated piping. (Use total actual footage of circulated piping for gpm calculation, but only the equivalent length of the longest run for the head calculation.)

Pipe size, in	gpm/100 ft[a]
1/2	0.29
3/4	0.30
1	0.33
1-1/4	0.37
1-1/2	0.40

[a]Based on limiting temperature rise to 5°F in a 90°F room.

(5) Size local cooler units, not an integral part of the dispensing unit, on the number of outlets (5 gph each) or, if known, the number of people (at 20 per gph) served, delivering water at 50°F.

(6) Provide filters on all central units. Provide a minimum of two (each 1/2 size) for each system (with as many as required for larger systems. Separately valve each filter.

(7) For developing equipment, consult the chilled-water unit manufacturer for his recommendation of the unit usually provided to supply a given piece of equipment.

(a) Locate the unit as close as possible to

        the equipment.
    (b) Provide oversized heavy-duty filters, to handle the continuous-flow requirements.

6. Distilled-water systems.

    a. Principles of design.

        (1) Provide distilled water in laboratories and/or at corridor dispensing stations in the laboratory areas, in pharmacies, and elsewhere, as required by the program. Review with the owner the expected quality of water (purity).
        (2) Where requirements are numerous in an area, provide a central system.
        (3) Where requirements are widely isolated, provide local units.
        (4) Locate central units in the penthouse and feed by gravity to outlets.

    b. Piping.

        (1) Consult the owner regarding the piping materials to be used, because some owners have very strong feelings about the matter. Possible materials are

            (a) Block tin-lined brass pipe (ID 1/8 in less due to lining)
            (b) Block tin-lined copper tubing (ID 1/8 in less due to lining)
            (c) Polypropylene pipe
            (d) Glass pipe
            (e) PVC pipe
            (f) Type 304-L or 316-L stainless-steel tubing
            (g) Aluminum pipe (Type 3003)

        (2) Avoid glass pipe, if possible, because it is fragile and should not be used when glass washers with solenoid valves are involved because it will not stand up under shock conditions.

    c. Equipment

        (1) Central units should be duplex (each one half

size) hospital-type, steam-heated completely automatic stills, and storage tank.
(2) Size the tank at five to ten times the hourly capacity of the stills.
(3) The following are the available tank materials and sizes:

   (a) Tin-lined copper (round vertical), 5 to 300 gal.
   (b) Tin-lined (steel or aluminum) rectangular, 100 to 1,500 gal.
   (c) Type 304 stainless-steel (round vertical) 10 to 1,000 gal, (round horizontal) 300 to 3,000 gal.

(4) Lacking detailed requirements from the owner, size stills on the basis of two people per 20 X 10 ft module, allow 1-1/2 gpd per person plus 25 percent wastage, plus 50 percent for growth.

   (a) Allow 1-1/2 gph per corridor outlet.
   (b) Select the stills to generate daily requirements in 20 h.
   (c) For glass washers, allow 45 gph (total 160 gpd).

(5) Note that stills can be operated on water pressures as low as 10 psi, and are not restricted to the 40 psi stated in their catalog. However, where less than 40 psi is used, automatic valves on the still must be specified as air-operated; and an air connection must be obtained from HVAC control air system. In all cases, the operating water pressure at the still should be stated in the specifications.

7. Demineralized (Deionized) water systems.

  a. Principles of design.

    (1) Provide demineralized water in laboratories and/or at corridor dispensing stations in the laboratory areas, and elsewhere, as required by the program. Review with owner the exact expected quality of water (purity).

(2) Where the requirements are numerous in an area, provide a central system.
(3) Where the requirements are few or isolated, provide local cartridge-type units.
(4) Locate the demineralizers in the basement with a pressure feed to the outlets.

b. Piping.

(1) Pipe materials: See Section C.6.

c. Equipment.

(1) The demineralizers should be completely automatic two-bed ion-exchange units with all required accessories, all selected for use on the raw water available and to produce water conforming to the owner's requirements.
(2) Factors for sizing demineralizers:

   (a) Allow 0.5 to 1 gpd per person for classroom laboratories (assume two classes per day).
   (b) Allow 1 to 1.5 gpd per person for other laboratories.
   (c) Allow 5 gpd for classroom preparation room outlets.
   (d) Plus wastage factor of 25 percent.
   (e) If feeding stills, add that load.
   (f) If feeding glass washers, add that load.
   (g) Allow 25 gpd for pipette washer.

(3) With the chemical content of the water and the gpd load known, the time between regenerations can be calculated. Tentatively, select equipment with between a minimum of a 7 day run between regenerations and an optimum of 2 weeks. Next, calculate the average peak flow rates of the system. If the tentatively selected equipment or equipment slightly larger can handle the expected flow rates, that is the selection. However, if the equipment to handle the flow rates is considerably larger, use tentative selected equipment plus a storage tank.
(4) If raw-water booster pumps are required, they can be normal IBBM units. If deionized water booster pumps are required, they must be all

stainless steel (required with tank systems). The water-distribution piping should be recirculated back to the tank by the booster pumps through a back pressure valve.
   (5) Tanks: See Section C.6.
   (6) Depending on the raw-water conditions, a carbon filter and/or a softner may be required ahead of the deionizer.

8. Reverse osmosis (RO) water systems.

   a. Principles of design.

      (1) Provide RO water in laboratories and/or corridor dispensing stations in the laboratory areas, and elsewhere, as required by the program. Review with owner the exact expected quality of water (purity).
      (2) Locate the units in the basement with a pressure feed to the outlets.

   b. Piping.

      (1) Pipe materials: See Section C.6.

   c. Equipment.

      (1) As the systems are specialized to the project requirements, after obtaining a complete raw-water analysis and the owner's requirements, consult with the manufacturer regarding equipment selection.

9. Water treatment.

   a. Principles of design.

      (1) Provide water treatment, where necessary, to furnish a piece of equipment with water of the quality essential for its operation.
      (2) Water treatment equipment should be of a type and size as required for the duty that it has to perform.

   b. Piping.

      (1) Limit copper and brass piping carrying completely softened water to a maximum velocity of 5 fps.

c. Equipment.

   (1) Water softeners.

      (a) Resin exchange units (usually with automatic regeneration) complete with brine tanks, interconnecting piping, and all required accessories and controls.

      (b) Normally units are sized on the gpm flow requirements of the system or equipment served. However, where gpm requirements are high compared with gph requirements (as for laundry washers), it may be more practical to base the size of the unit on the gph requirements (average gpm) and provide a storage tank to supply the peak gpm flows.

      (c) Hardness is expressed in grains per gallon as calcium carbonate equivalents.

      (d) Normal resins have a softening capacity of 30,000 grains/ft^3 when regenerated with 15 lb of salt per cubic foot of resin.

      (e) Softening capacity between regeneration equals the cubic foot of resin X 30,000 divided by the grains per gallon of the water softened.

      (f) Acceptable periods between regenerations depend on the operating schedule of system or equipment being served. Minimum should be 1 day.

      (g) If 100 percent continuous supply is required, multiple units must be provided to maintain supply during regeneration of one of the units.

      (h) Automatic regeneration can be actuated by the quantity of water used or by a time clock.

   (2) Filters.

      (a) The type of filter used depends on the filtration job it has to do and the required quality of the effluent.

      (b) Filters may be sand, diatomite, cartridge, or carbon type as required by (a) above.

10. Nonpotable-water systems.

   a. Principles of design.

   (1) Where required by code, completely separate by an air gap the water piping serving flushing rim floor drains, bidets, mortuary equipment, autopsy equipment and other code required equipment from the other building water piping.
   (2) Use an over-the-rim supplied surge tank above, with gravity feed; or where this is not possible, use an over-the-rim supplied surge tank with a booster pump or a hydropneumatic pressure system to accomplish this.

   b. Piping.

   (1) Piping materials: See Section C.1.

   c. Equipment.

   (1) Equipment: See Section C.3 and C.4.

D. Irrigation systems.

   1. Principles of design.

      a. Provide irrigation systems where requested by the owner and/or the architect.
      b. Determine from the architect the exact areas involved, and the type of planting, i.e., grass, flowers, shrubbery, trees, etc., in each area.
      c. Provide the water supply to irrigation systems with approved backflow preventers. Check the code for detailed local requirements.
      d. Various types of plant materials require varying amounts of water and different types of heads; consequently a single zone of sprinklers should not be used to accommodate two different types of plant material (e.g., turf and shrub beds). Select the proper type of head for the growth it is to serve, as recommended by the head manufacturer.
      e. Types of heads available.

         (1) Spray type: Pop-up or stationary. Best for lawns. Operates on relatively low pressure. Triangular spacing preferred, using 60

percent of diameter of throw (generally up to 15 ft radius).

(2) Bed spray heads: For shrubs, flower beds, and ground cover. Installed on risers either above or below the foliage.

(3) Strip or line type: Stationary for narrow areas and edges.

(4) Bubbler heads: For soaking small areas. Installed above ground. Flood type for very small areas and stream type for slightly larger areas, and under extremely dense foliage gives low water rate.

(5) Rotary type: For large areas, has long throw, low water rate, and requires higher pressures. Good for steep slopes. Hard to control in wind. Square pattern preferred using 60 percent of diameter of throw.

   (a) Gear drive: Pop-up. Small 16 to 30 ft radius, and large 30 to 80 ft radius. Valve in head type over 80 ft radius.
   (b) Impact drive: Installed above ground or pop-up over 80 ft radius.
   (c) Cam drive: Pop-up. Troublesome if water is not clean.

(6) For athletic fields, provide special rubber covered heads.

(7) Risers for above ground heads should be adjustable type.

f. Uses of various typeheads.

(1) Spray heads (pop-up or stationary) are generally used in residential or small areas where large diameters are not required or where water quantities are limited. Spray coverage is excellent and uniform.

(2) Rotary heads.

   (a) Gear drive (pop-up), medium-sized installation requiring greater radius than 15 ft.
   (b) Impact type heads (pop-up or fixed) are used for large installations such as large commercial sites, golf courses, or cemeteries. These heads have large areas of coverage and use high quantities of

water at good pressures. Impact heads do not give as even a spray coverage as other types of heads.

(3) Select special type heads for trees, shrubbery, strips, edges, etc.

(4) Try to use as many 360° heads as possible in a given system since the costs for installation of full-circle and part-circle heads are approximately the same while the area covered is much less. The use of an excessive number of part-circle heads will thus result in a higher unit price per area sprinkled.

(5) If part-circle heads are used, check to see if the watering density is different from the full-circle heads. If there is a substantial difference, the part-circle heads should be valved on a separate zone to permit the varying of the time cycle to avoid overwatering or underwatering. Varying densities with part-circle heads may occur with any type of spray head.

(6) Overlap of sprinkler patterns is required. Layouts of sprinklers where spray patterns are tangent are incorrect. In laying out the system, the manufacturer's recommendations for spacing are to be followed; not the criteria for spray radius. Radius criteria is not to be used, since this figure refers to effective radius in some catalogs and extreme radius in others.

(7) In areas which are confined by sidewalks, buildings, parking lots, etc., provide 180° heads at the edge of these areas which throw into the area to be irrigated. Do not locate heads which could wet areas not intended to be watered.

(8) In areas where wind is anticipated as a problem, the following corrections should be made:

(a) Where wind is in a constant direction, shift sprinklers into the wind by 1/2 to 1 ft per mph of wind velocity. (Use average wind velocities during the watering season, not extreme values.)

(b) Where wind direction changes, reduce spacing (on triangular basis) by about 10 percent per 10 mph of wind velocity.

(c) For part-circle heads, increase or decrease the part-circular coverage as required by examination to give even coverage.
(d) For heads along walks, roads, etc., where wind is a factor, use 200° heads instead of 180° heads when wind is blowing away from walks and 140° heads instead of 180° heads when wind is blowing towards walks.
(e) On shallow slopes or in valleys, use the actual surface dimensions and area for calculation, not the plan dimensions (compensate for slope). Heads may be tilted slightly to give optimum spray coverage. However, this tilting and adjustment should be made in the field. On steep slopes, tilt head slightly and cut back the upstream side slope. Spacing on steep slopes should be approximately the same (or perhaps a bit tighter) as for normal areas.

g. Sections should be fed as near the center as practical to minimize pipe sizes.

h. Automatic sectional valves should be selected with or without pressure regulators as required.

   (1) Valves should be installed inside the building when ever practical.
   (2) Underground valves should be provided with plastic valve boxes.
   (3) All automatic valves should be preceded by control valves.
   (4) Automatic valves should be carefully selected for the required flow, often smaller than line size.

i. When manual sectional valves are used, they should be located so that the operator will not be wet by the spray, and they should be easily located.
j. Each section (zone) should be pitched to and provided with an automatic drain valve.
k. Minimum cover on piping should be 18 in.
l. Heads in areas subject to heavy equipment travel over them should have swing-type connections.
m. Irrigation systems should be provided with automatic controllers to properly time the watering

of each section and to stagger the operation of the various sections to minimize the required peak water demand (flow). Controllers should be the heavy-duty type.

    n. Where available pressure is insufficient for the requirements of a properly designed system, provide a booster pump. Booster pumps should be sized on the estimated peak flow required and the pressure differential required to obtain the required pressure at the farthest head.

    o. As watering is usually done in off hours, its demand normally is not added to the peak domestic demand on the project.

2. Pipe sizing.

    a. Size piping based on the estimated peak flow in that particular piece of pipe, giving due consideration to the sectionalizing of the whole system.

    b. Calculate friction losses based on the appropriate friction tables for the type of pipe being specified. Where an option of pipe materials is specified, use the friction table of the acceptable pipe with the highest friction losses.

3. Piping.

    a. Piping inside the building should be as for domestic water supply systems.

    b. Underground piping should be cast-iron, ductile-iron, asbestos cement (Transite), or plastic of proper pressure rating: polyvinyl chloride (PVC) or polypropylene (PP) with socket welded fittings, or polyethylene (PE) with banded insert fittings. Except for large volume systems (large pipe sizes), generally plastic will be the pipe material used.

    c. Plastic piping can be specified using the SDR pressure designations. This allows specifying a greater variety of pressure ranges depending on the specific job conditions and enables incorporation of economies in the system.

    d. Control valves on plastic piping should be plastic and provided with plastic extension boxes to grade.

    e. Where automatic control valves are located out in the system rather than in the building, pressure rating of plastic piping ahead of valves should be selected generously to allow a safety factor for water hammer.

E. Swimming pools.

1. Principles of design.

   a. Swimming pools, except for private residential pools (and in some localities, private pools) are governed by code.

      (1) Usually the Department of Health code, rather than the plumbing code (New York City-both) is followed.
      (2) Design swimming pools in accordance with the requirements of all local authorities and the National Swimming Pool Institute, whichever is the more restrictive.
      (3) The requirements usually govern the following:

         (a) Time of turnover (recirculating rate); usually 6 to 8 h.
         (b) Types of filters and filtration rates.
         (c) Spacing of main drains and grate area requirements.
         (d) Spacing of inlets.
         (e) Spacing and capacity of gutter drains and surge tank or skimmers (usually outdoor pools only).
         (f) Heating requirements.
         (g) Chlorination requirements.
         (h) Draining requirements (both time and to which sewer system).
         (i) Fill and makeup requirements.

   b. Consult your supervisor about the design of all swimming pools.
   c. Types of pools are as follows:

      (1) Wading pool: maximum water depth 2 ft.
      (2) Swimming pool: depth over 2 ft.
      (3) Hydrotherapeutic pool: a swimming pool (usually approximately 95°F) in a hospital used for treating patients.

   d. Where no code exists or an item is not covered in the code, use the fillowing design criteria:

      (1) Recirculation rate.

         (a) Swimming pools; 6 to 8 h turnover.

Lightly loaded pools can use 8 h, but heavily loaded pools should use 6 h.
(b) Hydrotherapeutic pools should use 4 to 6 h.
(c) Wading pools should use 4 h.

e. Main drains should be sized for 100 percent of the recirculation rate with grates having an area of at least 4 times the area of the outlet pipe or an area that produces a maximum velocity of 1-1/2 ft/s under maximum flow conditions. The drains should be located a maximum of 15 ft from the side wall with a maximum of 30 ft between drains.

f. Gutter drainage system should be sized for 50 to 100 percent of the recirculating rate with a minimum 2 in drain outlet. Drains should be spaced a maximum of 15 ft apart. Grate area should be at least 1-1/2 times the area of the outlet pipe. Special patented continuous-flow gutters (with adjusting weirs) and supply assemblies are sometimes requested. Check with local code authorities about its acceptability.

g. If used instead of gutters, skimmers should be provided for every 400 to 500 ft^2 of pool area and sized for 80 to 100 percent of the recirculation rate with a minimum of 30 gpm per skimmer, or if over 8 in wide, 3.75 gpm per inch of weir length. Provide skimmers with equalizer pipes. Some model skimmers also include an integral vacuum cleaning connection, thus eliminating the need for separate vacuum cleaning connections on the pool wall.

h. Provide surge (balancing) tanks for pools (except private pools) using sand filters, unless skimmers or patented continuous-flow gutters are used. Allow surge capacity of 1/2 to 1 gal/ft^2 of pool area. Surge tanks are not usually required with vacuum diatomite filters, skimmers, or patented continuous-flow gutters.

i. Locate inlets around the perimeter of the pool to give an even distribution. Maximum spacing 20 ft and with an inlet within 5 ft of a corner.

(1) Inlets should have an integral throttling device to balance the system.
(2) Sometimes pools (mainly those used primarily for diving) are designed with their inlets in the bottom, using the gutter (or skimmers) for recirculation, and the main drain only

for emptying. If underground, this has the bad feature of placing the supply piping under the pool where it is very costly to get at for repairs.

j. Vacuum cleaning outlets should be spaced as required to cover the pool using not over 50 ft of hose. The size and type of cleaning tool should be selected based on the size and shape of the pool.
k. Chemical treatment equipment, hypochlorinators, should maintain a free chlorine residual of 0.6 to 1 ppm with a Ph of 7.2 to 7.8.
l. Pool water temperatures should be 76°F to 78°F.
m. Fill and makeup connections should have an air gap, either over the rim of the pool, the surge tank, the vacuum diatomite filter, or the makeup tank unless the code permits the use of an approved reduced-pressure back-flow preventer.

   (1) Makeup connections should be provided with a solenoid valve or solenoid pilot controlled valve, interlocked with the high-water alarm and the pump to prevent overflow or flooding if pump is not running.
   (2) When makeup water is introduced through an air gap, provide an open- and closed-type (modulating type is not necessary) float valve with speed controls.
   (3) Provide a full-size valved line for filling and a smaller-size bypass including float valve and throttling valve for makeup.

n. Provide all pool recirculating systems with an approved rate-of-flow indicator. New York City also requires that it be a recording chart unit.
o. Provide pressure gauges in the discharge of all recirculating pumps, the suction of all recirculating pumps, and the inlets of all strainers. Check whether the pump suction and strainer inlet gauge should be compound type.
p. Provide pools with drains in the deck surrounding the pool. Pipe these deck drains to the sanitary system for inside pools and the storm system for outdoor pools.
q. Provide underground pools (except small residential) with a tunnel around them to carry the piping and make it accessible, rather than burying

the piping.
r. Provide underground pools, dipping into ground water, with hydrostatic relief valves in the bottom.
s. Carefully select all pool fittings for their purpose and location.
   (1) In selection, special attention must be paid to their proper connection to the architect's waterproofing system. Membrane requires lead flashing "burned" to the body (bronze) (not clamped) and elastomeric and liquid waterproofing requires a wide flange.
   (2) Drains must be without weep holes. If the drain is selected from the swimming pool section of the catalog, it has no weep holes. If it is selected from the roof drain or floor drains sections, it must be noted to be without weep holes. Carefully check this point on shop drawings.
t. Supply piping should have a maximum velocity of 8 fps. Recirculating return piping to pumps should have a maximum velocity of 5 to 6 fps.
u. When an aluminum pool is involved, all possible copper or copper alloys must be eliminated from the system.

   (1) In addition, an easily replaceable sacrificial piece of aluminum piping at least 2 ft long and with at least one elbow (the more the better) and having flanged or union ends should be provided in every line carrying water to the pool.
   (2) Provide insulating bushings, couplings, unions or flanges whenever any other metal contacts the aluminum.

v. See Section V for insulation requirements.

2. Piping.

   a. Fill and makeup water piping shall be as specified for domestic water piping.
   b. As recirculated pool water is more corrosive than domestic water, it requires piping materials as good or better than that required for the domestic water.

      (1) Inside the building, small piping should be

4-112

copper tubing or red brass piping. Larger sizes should be bitumastic enamel (AWWA specifications), epoxy, or plastic-lined steel pipe and malleable or cast-iron fittings.

(2) Underground piping may be polyvinyl chloride (PVC) or polypropylene (PP) pipe and fittings.

    (a) In small sizes it may be copper tubing.
    (b) In larger sizes it may be cement-lined cast-iron or ductile-iron bell and spigot pressure pipe and fittings or asbestos cement (Transite) pressure pipe with cast-iron or ductile-iron fittings.

(3) As most control valves are used for throttling, they should be ball valves in the smaller sizes and butterfly valves in the larger sizes.

c. Inside the building, plastic pipe and fittings may be used, if approved by your supervisor. Where piping is insulated, fire and smoke ratings of the insulation supercedes those of the pipe and fittings.

3. Equipment.

  a. Filters.

    (1) Rapid sand filters (pressure type): Rate 3 gpm/ft^2. Require at least three units as they must be backwashed at 3 to 4 times the recirculation rate. They are wasteful of water and require the most space. Require alum and soda ash feeders to operate.

    (2) Vacuum diatomite filters: Rate 2 gpm/ft^2. Require little water for backwash and minimum space. Due to fiberglass tank, they require no painting. The open tank can serve as the required air break and eliminate the need for a makeup, surge, or balancing tank.

        (a) They require a body coat feeder.
        (b) They require a pit to collect the used diatomite for manual removal. This is their major disadvantage.
        (c) Precoating and backwashing can be automated but is expensive.

(d) Where a vacuum-type filter is used, it is imperative that an accurate section be drawn showing the elevations of the pool, the filter, and the filter room, because these vertical distances are critical in the selection of a filter and its control and must be made known to all manufacturers bidding on the job. In making these sections, consideration should be given to the foundation under the filter, as some filters require certain minimum height foundations due to their construction.

(3) High rate sand filters: Rate, 15 gpm/ft^2. Backwashed at recirculating rate.

(a) The latest type therefore not yet recognized by some codes. May be allowed upon special request. So far, mainly used in the small sizes.

b. Pumps.

(1) The recirculating pumps should be flexible coupled horizontally split case, flexible coupled, or close-coupled vertically split case, depending on the size. With sand filter systems, it is sometimes possible to use the recirculating pump as the vacuum cleaning pump by sliding it on its curve, if it is too large, or by providing two half-size recirculating pumps.

(2) A vacuum cleaning pump must be provided on sand filter systems where the recirculating pump (or pumps) cannot meet the vacuum system requirements, and on vacuum diatomite filter systems. Pump should be close-coupled vertically split case or flexible coupled horizontally split case, depending on the size.

(3) The head on the pump in a bottom recirculating system with a pressure filter and no surge tank should be the sum of the following: The pressure required at the outlet (for patented continuous-flow gutters allow 10 psi) converted from psi to feet, the discharge friction, the loss through the heater,

the loss through the filter, the suction friction, and the loss through the strainer.

(4) The head on the pump with a surge tank or vacuum filter should be the sum of the following: The pressure required at the outlet (for patented continuous-flow gutters allow 10 psi) converted from psi to feet, the lift from the pump to the pool water level, the discharge friction, the loss through the filter converted from psi to feet, the friction in the suction piping from the tank, and the loss through the strainer converted from psi to feet.

    (a) For vacuum diatomite filters, allow 20 ft for the loss through the filter.
    (b) For high-rate sand filters, allow 15 to 20 psi converted from psi to feet for the loss through the filter.

(5) NPSH requirements of pumps should be carefully checked.
(6) Pool pumps are manually operated.
(7) Pumps drawing from tanks should be provided with low-water cutoff switches.

    (a) Except for recirculating pumps pulling through diatomite filters, all recirculating pumps should be provided with full line size quick-opening basket strainers in their suction piping.

c. Heater.

(1) The heater should be a shell and tube heat exchanger. The heater must be able to heat the water on filling the pool and heat the required makeup. As these two requirements are widely different, the sizing must be a compromise, requiring more than one turnover to bring the freshly filled pool up to temperature, but not too long.
(2) In small sizes, usually all the water is passed through the heater with the required temperature rise.
(3) In large pools, it is usually more economical to pass only part of the water through the heater at an increased temperature rise, then

blend it back with the bypassed water.
(4) Advise the HVAC project engineer of the steam requirements.

d. Sterilization equipment.

(1) Sterilization of the pool water to kill bacteria, and in addition for outdoor pools, to eliminate algae, is accomplished by the use of a hypochlorinator (chemical feeder).

   (a) Hypochlorinators in public pools are usually required to be provided in duplex.
   (b) This equipment may be automated.

(2) Automation requires an additional chemical feeder for pH control.

(3) The automation equipment gives direct reading of the chlorine residual and the pH of the pool water. It can also be bought to produce continuous reading charts, if the owner so desires.

(4) Hypochlorinators should be small, adjustable, positive displacement pumps (chemical feeders), built specially for the service, drawing sodium hypochlorite from a container and injecting it into the filter discharge piping to the pool after the last (hot-water heater) connection.

(5) Hypochlorinators must be provided with plastic containers to hold the required sodium hypochlorite, and the acid and/or alkali required for pH control.

   (a) Check that the units selected have a discharge pressure capability greater than that in the piping that they connect to.
   (b) In the table are suggested minimum sized hypochlorinators:

System volume, gal	Suggested size, gph[a]
10,000	0.14
18,000	0.25
42,000	0.58
60,000	0.83
72,000	1.0
114,000	1.6
180,000	2.5
360,000	5.
720,000	10.
972,000	13.5
1,300,000	18.

[a]Capacity at maximum adjustment rating.

    (6) Hypochlorinators should be interlocked with the filter recirculating pump so that they operate only when the pump is running.

  e. In the table are the requirements of the various sizes of vacuum cleaning tools:

Size	Hose, in	gpm
12-in hand	1 1/4	35
15-in hand	1 1/2	50-60
18-in hand	2	70-80
24-in tow	2	80-100
30-in tow	2	100-120

    (1) Use tow model in pools over 30 ft wide.
    (2) Allow 32 ft of pump head for the tool plus friction in the hose, piping, strainer, etc.

F. Decorative pools and fountains.

  1. Principles of design.

a. Determine from the architect exactly what is wanted in the way of a display, with respect to

   (1) Number of nozzles and height of sprays
   (2) Type of nozzles
   (3) Special effects such as water falls
   (4) Winterization
   (5) Wind control

b. In conversations with the architect, point out the following: Vertical nozzles on a normal (not excessively windy day) can wet an area 1 to 1-1/2 times the spray height. Usually architects will not believe this. If you cannot convince them, design the system as they desire, and it will be throttled down to usable conditions after installation.

c. Water falls splash when they hit the water surface in the base pool; therefore, an adequate space must be allowed in the pool in front of a water fall (20-ft water fall requires at least 8 ft).

d. Generally decorative display fountains and pools are not covered by code (New York City does have code requirements) except for fill and makeup water connections.

   (1) Your supervisor should be consulted on the design of all display fountains and reflecting pools.
   (2) Decorative display fountains and pools should be designed on the basis of good engineering (hydraulics).

e. Except for very small fountains, all decorative fountains and pools should have a filtered recirculating system independent of the nozzle recirculating systems, because it must operate 24 h a day while the nozzles do not. The turnover rate should be approximately 12 h. In a winterized installation, this may have to be increased to keep temperature of the return water from being too high, to prevent steaming of the surface.

f. Because the architect is often not sure about what he wants to see, the pump and nozzles should be selected to allow for some upward adjustment after installation and the architect actually sees what it looks like. This is not necessary if the architect's height requirements are greater than is practical to operate the fountain.

g. A fountain is never fully designed in the design stage; the design is finally completed by adjustments in the field after installation.

h. In an unusual or complicated setup, it may be necessary to have a mockup or witness a test at the manufacturer's plant before completing contract documents.

i. To allow for future adjustments, an adequate number of throttling valves and rate-of-flow indicators should be provided, as well as a valved bypass between all jet recirculating pump discharges and suctions.

j. The pool water depth must be deep enough to allow for the proper installation of the size and type of nozzles selected.

   (1) Aerating jets require greater water depths than straight jets.
   (2) In New York City, the maximum allowable water depth is 18 in.

k. Before finalizing the pool water depths with the architect, consult with the electrical project engineer, because his lights may require minimum water depth that may govern in some cases. To save on weight of water in the pool, sometimes only the area immediately around the cluster of nozzles and lights is depressed to give the required water depth without increasing the depth of the entire pool.

l. In considering the free-board volume of a gutter or lower pool receiving the flow from a water fall, remember that when the installation is operating, there is a measurable layer of water above the waterfall weir over the entire pool, which, when the pump stops, must be stored in the gutter or lower pool until the pump starts again. If space permits, it is also desirable to include provision for some storage of rain water to save on makeup water costs.

m. Nozzles are usually bronze and are available in many different sizes and shapes to give a wide variation of display patterns. However, all nozzles are basically one of two types: straight jets or aerating jets.

   (1) Straight jets, which give a solid stream of

water.

    (a) They are best for long throws and the only type available for flat or fan stream or adjustable mushroom displays.
    (b) They are the only type that will operate completely out of water.
    (c) Mushroom nozzles may be obtained with a vertical jet in the center. This type of nozzle has two inlet connections each connected to a separate system of piping due to the difference in pressure required for each display.

(2) Aerating jets, which pass the pumped water through a throat sucking in air and sucking up additional water from the pool.

    (a) Since they do not require the pumping of all the water thrown into the air, the pumps and piping are smaller; however, due to the higher pressures required for the air and water pickup, horsepower may be greater.
    (b) Depending on the model selected, they can pick up 2 to 4 times their pumped flow from the pool.
    (c) Aerating jets produce a bushy foamy column that sparkles better under the lights than does the solid stream of a straight jet.

(3) Certain nozzles of the aerating-jet type are water-level-dependent and require close plus and minus water-level tolerances for proper operation.

    (a) Pools with waterfalls from them automatically take care of this problem. However, the nozzles must be installed with the tops related to the water level when in operation (weir height plus the thickness of the sheet of water going over the weir).
    (b) In other designs, automatic water-level controls (controlling makeup) are essential, rather than the manual makeup usually provided.

    (c) Some models of aerating jets are adjustable, allowing a variation of the amounts of water and entrained air, which changes its appearance.
    (d) Some models of aerating jets require suction screens around their pool suction inlets.

(4) Check the manufacturer's catalog for recommendations on the need for a splash screen with the type of nozzle selected.

(5) To achieve the best possible appearance, most nozzles require linear, nontwisting flow into the nozzle, which means that there should be a straight section of pipe between the nozzle and a fitting and/or valve ahead of it. Where it is impossible to install a sufficient length of straight pipe ahead of the nozzle, flow vanes should be installed in the pipe ahead of the nozzle or results produced by the nozzle, as installed, accepted. Check with the manufacturer for recommendations when the above is not possible.

(6) For nozzles installed other than vertical, see Figure 4-9 for conversion to vertical height for selecting capacity requirements.

(7) All nozzles should be provided with ball joints beneath them to ensure that the rising water is absolutely vertical regardless of the workmanship.

n. Waterfalls, another type of water display, use weirs instead of nozzles to control the flow.

(1) Where the waterfall starts the display, a trough of water must be provided behind the weir to produce a constant even flow of water approaching the weir.

(2) Weirs should be provided with adjustable vertical bronze or stainless-steel weir plates to allow for even level adjustments, because it is impossible to construct masonry absolutely level.

(3) Following are some suggested flow rates over weirs. These are subject to modification to suit a particular project and the architect's desires.

(a) The higher the waterfall, the greater the flow must be to look good.

(b) With short weirs, generous flow rates can be used because the short footage does not give excessive total flow. However, with long weirs, flow rates must be conservative, because the long footage gives excessively high total flows, which can be costly in pipe size, pump size, and operating cost.

gpm/ft	Thickness of water over weir, in	Approximate normal maximum height of waterfall, ft
15	1/4 (0.3)	1
20	3/8 (0.4)	3
25	1/2 (0.5)	5
30	5/8 (0.7)	7
35	3/4 (0.8)	10
40	7/8 (0.9)	15
50	1 (1.0)	25
60	1-1/4 (1.3)	Over 25

o. Where nozzles requiring widely different pressures are involved, a decision must be made about the use of one recirculating system with pressure-reducing or throttling valves, or multirecirculating supply systems. (Return can be common from the same pool.) Generally, where the lower pressure is considerably lower or the quantity involved is much greater, multicirculating systems should be used to conserve energy.

p. Pool inlets in the filtration, chlorination, and heating recirculating system should be of sufficient number and carefully located around the pool and with respect to the drains, to ensure an even distribution of the chlorinated and heated water throughout the pool.

q. Pools are usually recirculated through the bottom drain. Occasionally due to neighborhood conditions (trees, shrubs, etc.), the architect or owner may request that recirculation be via a gutter or skimmers. Do not use skimmers where the water level is critical to the nozzle operation.

(1) Pools recirculated through the bottom drain are connected directly to the pump through a large strainer (no tank required).

(2) Pools recirculated through a gutter or skimmer require a tank in the equipment room to store sufficient water when the recirculating pump is shut down, to keep the pump supplied with water at startup until the first water pumped returns from the pool. In addition, tanks should be sized, if space is available, to store some rain water (free makeup water).

(3) Normal fill and makeup would be introduced into this tank, and must be automatically controlled at both pump-running and pump-off levels in the tank.

r. Recirculating outlets in the pool bottom should be provided with adjustable Anti-Vortex plates. This consists of a plate at least 4 in larger than the drain body, supported on four long threaded stud bolts brazed to the drain body.

s. Where a waterfall is involved and the nozzles do not supply sufficient water for a proper waterfall, additional water (a separate recirculating supply system) must be provided to the pool under water without breaking the surface.

(1) The inlet fitting for this is similar in type to the outlet drains described above, except the Anti-Vortex plate is called a Suppression plate.

(2) Inlet fittings to the trough feeding waterfalls should be provided with a Suppression plate, as described above.

t. All pools and fountains should be provided with overflows to prevent flooding in the event of abnormal rainfall or makeup valves stuck or inadvertently left open. For pools without tanks, overflows should be either freestanding (with dome) or wall type in the pool, as per architects desires and/or construction requirements.

u. All pool fittings should be carefully selected for their purpose and location.

(1) In selection, special attention must be paid to their proper connection to the architect's waterproofing system. Membrane requires lead

flashing "burned" (not clamped) to the body (bronze); and elastomeric and liquid waterproofing requires a wide flange.

(2) Drains must be without weep holes. If the drain is selected from the swimming-pool section of the catalog, it has no weep holes; if selected from the roof-drain or floor-drain sections, it must be noted to be without weep holes. Carefully check this point on shop drawings.

(3) For piping through waterproofing, use the waterproofing attached sleeve as a coupling in the piping.

v. If the pool is to be winterized, consult with the HVAC project engineer about recommended heating requirements.

(1) Sometimes all heat is added in the filtration recirculating system and sometimes heat is also added in the jet system.

(2) Give the project engineer your desired filtration rate (gpm). He may have to ask you to increase it if your rate would result in too high a temperature of the water returning to the pool (which would cause steaming on the surface of the pool on a cold day).

w. Normally all nozzles should be provided with individual throttling valves to even out the spray heights regardless of varying friction losses to the individual nozzles and for overall height adjustment.

x. When the nozzles are few and/or spread widely apart, the supply to each should come up through the bottom of the pool separately. Where the nozzles are closely spaced, provide a header or ring in the pool to feed the nozzles, with supply feeds as required through the bottom of the pool, to minimize penetrations of the waterproofing. Note that headers or rings in the pool usually increase the minimum required water depth.

y. Fill and makeup connections should have an air gap, unless the code permits the use of an approved reduced-pressure back-flow preventer.

(1) Makeup connections should be provided with a solenoid valve or solenoid pilot controlled

valve, interlocked with the jet pump to prevent overflow or flooding during nonoperating periods.

 (2) When makeup water is introduced through an air gap, provide an open- and closed-type (modulating type is not necessary) float valve with speed controls.

 (3) Provide a full-size valved line for filling and a smaller size bypass including float valve and throttling valve for makeup.

z. To facilitate adjustments, the following should be provided:

 (1) Rate-of-flow indicators: In the discharge of all recirculating pumps, the makeup line and all jet subsystems.

 (2) Pressure gauges: In the discharge of all recirculating pumps, all jet subsystems, the suction of all recirculating pumps, and the inlets of all strainers. Check whether the pump suction and strainer inlet gauges should be compound type.

aa. Wind controls may be one- or two-stage, to reduce the fountain height (in one or two stages) or to shut the fountain down completely.

 (1) To lower the height of the jets, the wind control either closes an automatically controlled throttling valve in the pump discharge a set amount or opens an automatically controlled throttling valve in the discharge suction bypass a set amount or both.

 (2) To shut the fountain down completely, the wind control opens fully an automatically controlled throttling valve in the discharge suction bypass or shuts the pump off. In the latter case, a time-delay must be provided to prevent too-frequent operation of the pump.

 (3) Automatically controlled throttling valves should be pneumatically operated through a PE switch.

bb. The filtration recirculation system should be provided with a hypochlorinator to prevent algae growth in the pool. For small pools and indoor pools, chlorination can be done by hand instead of mechanically.

cc. Supply piping should have a maximum velocity of 8 fps. Recirculating return piping to pumps should have a maximum velocity of 5 to 6 fps.
dd. Decorative fountain and pool vacuum cleaning is usually done with a portable electric or gasoline engine driven pump rather than a piped central system as with a swimming pool.

   (1) Be sure to locate plaza drains of sufficient size in the general pool area to receive the discharge from the portable vacuum-cleaning pump.
   (2) Remind the electrical project engineer to provide waterproof electric outlets in the general pool area to operate a portable vacuum-cleaning pump.
   (3) The portable vacuum-cleaning pump, reel of wire (if electric), and cleaning tools are usually supplied by the owner after occupancy; however, they can be supplied by the plumber if the owner desires. Where a portable electric vacuum-cleaning pump is provided, consideration must be given to the reel of extension wiring. Especially if the wiring is long, the reel becomes quite heavy and will need to be provided with wheels as is the pump. It is desirable to limit the wiring on the reel to a maximum of 125 ft to avoid a large reel size.

ee. When an aluminum pool is involved, all possible copper or copper alloys must be eliminated from the system.

   (1) In addition, an easily replaceable sacrificial piece of aluminum piping at least 2 ft long, with at least one elbow (the more the better) and flanged or union ends should be provided in every line carrying water _to_ the pool.
   (2) Insulating bushings, couplings, unions, or flanges should be provided whenever any other metal contacts the aluminum.

ff. See Section V for insulation requirements.

Clearstream jet

Spray design calculations

Design Factors

A	E	C percent	D percent
5°	0.90	6	36
15°	1.33	11	46
25°	1.83	17	49
35°	1.94	22	51
45°	2.10	27	52
55°	1.80	36	53
65°	1.50	50	56
75°	0.90	99	59
85°	0.40	245	64

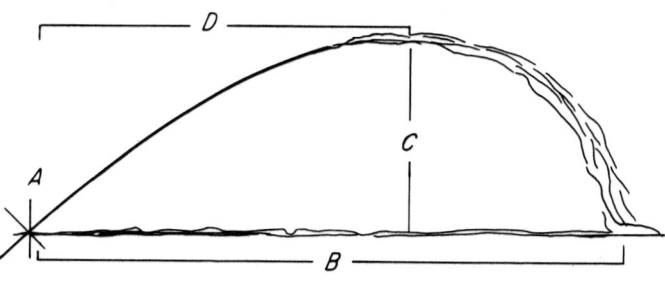

A = angle of nozzle elevation
B = horizontal distance of throw from nozzle.
C = height of trajectory in percentage of B.
D = highest point of trajectory in horizontal distance from nozzle measured in percentage of B
E = multiplying or dividing factor for spray calculation
X = vertical spray height

To find:	How:
1. Horizontal distance of throw for a desired angle of spray when only the vertical spray-height is known.	1. Establish vertical spray-height factor X and multiply the same by factor E to achieve horizontal spray distance.
2. Performance requirements of spray-patterns other than vertical or the equivalent vertical spray-height performance requirements.	2. Establish horizontal distance of throw from nozzle and divide by factor E. This will give vertical spray-height, which is used to find performance requirements.
3. Trajectory of stream or spray.	3. Establish B, then calculate factors D and C thereof; combine the results with B to lay out the trajectory.
4. The jet elevation angle (factor A) for the specification of particular trajectories or spray effects.	4. Establish B, calculate factor C thereof, and read on the factors table the value of factor A as found on the same line as the result of the calculated factor C.

All spray design calculations are based on linear, nonturbulent flow or water into the jet having minimum directional adjustment and no wind. Where turbulence is present and best possible performance is desired, the use of flow-straightening vanes in the pipe leading to the jet, at a distance of at least ten times the diameter of the pipe from the jet, is recommended. Turbulence and twisting, nonlinear flow will cause early breakup and reduced spray-height and spray distance.

Fig. 4-9. Angle spray design factors. Information supplied by PEM Fountain Co., Richmond Hill, Canada.

2. Piping.

   a. Fill and makeup water piping should be as specified for domestic water piping.
   b. Since recirculated pool water is more corrosive than domestic water, it requires piping materials as good or better than that required for the domestic water.

      (1) Inside the building, small piping should be copper tubing or red brass piping. Larger sizes should be bitumastic enamel (AWWA Specifications), epoxy, or plastic-lined steel pipe and malleable or cast-iron fittings.
      (2) Underground piping may be polyvinyl chloride (PVC) or polypropylene (PP) pipe and fittings.

         (a) In small sizes, it may be copper tubing.
         (b) In larger sizes, it may be cement-lined cast-iron or ductile-iron bell and spigot pressure pipe and fittings or asbestos cement (Transite) pressure pipe with cast-iron or ductile-iron fittings.

      (3) Since most control valves are used for throttling, they should be ball valves in the smaller sizes and butterfly valves in the larger sizes.

   c. Inside the building, plastic pipe and fittings may be used, if approved by your supervisor. Where piping is insulated, fire and smoke ratings of the insulation supercedes those of the pipe and fittings.

3. Equipment.

   a. Filters.

      (1) High-rate sand filters. Rate 15 gpm/ft^2. Backwashed at recirculating rate. The type to use unless the size required is large.
      (2) Vacuum diatomite filters. Rate 2 gpm/ft^2. Require little water for backwash and minimum space. Due to fiberglass tank, they require no painting. The open tank can serve as the required air break and eliminate the need for a makeup tank.

(a) They require a body coat feeder.
(b) They require a pit to collect the used diatomite for manual removal. This is their major disadvantage.
(c) Where a vacuum-type filter is used, it is imperative that an accurate section be drawn showing the elevations of the pool, the filter, and the filter room, since these vertical distances are critical in the selection of a filter and its control and must be made known to all manufacturers bidding on the job. In making these sections, consideration should be given to the foundation under the filter, because some filters require certain minimum height foundations due to their construction.

b. Pumps.

(1) The recirculating pumps should be flexible coupled horizontally split case, or close coupled vertically split case depending on the size. Where the local codes permit such an installation (not New York City) and the equipment room is rather remote from the pool, consideration should be given, as a matter of economy, to using a submersible pump (a deep well pump lying horizontal with an Anti-Vortex plate over it) in a depression in the bottom of the pool.

(2) The head on the pump in a bottom recirculating system with a pressure filter should be the sum of the following: The highest pressure required at the nozzle (or outlet) converted from psi to feet; the lift from the pool water level to the highest nozzle or trough behind a weir (if applicable); the discharge friction; the loss through the heater (if applicable); the loss through the filter (if applicable); the suction friction; and the loss through the strainer.

(3) The head on the pump in a system with a storage tank or vacuum filter should be the sum of the following: The highest pressure required at the nozzle (or outlet) converted from psi to feet; the lift from the pump to the pool water level, to the trough behind a weir, or to the highest nozzle, whichever is

applicable; the discharge friction; the loss through the filter (if applicable); the friction in the suction piping from the tank and the loss through the strainer (if applicable) converted from psi to feet.

   (a) For vacuum diatomite filters, allow 20 ft for the loss through the filter.
   (b) For high-rate sand filters, allow 15 to 20 psi converted from psi to feet for the loss through the filter.

(4) The initial selection of the jet-supply recirculating pump should be based on the required gpm and the head as calculated above. The final selection (and specification) should be based on the maximum that the required pump and motor can produce (to allow flexibility for field adjustment of fountain height); or if initial selection is the maximum that the pump and motor can produce, the next larger size pump and/or motor should be specified.

(5) NPSH requirements of recirculating pumps should be carefully checked.

(6) Normally fountain and pool recirculating pumps are manually operated, sometimes with additional start-stop buttons at a remote location.

   (a) Occasionally the jet-supply recirculating pumps may be requested to be provided with time clock or wind control.
   (b) As the filter recirculating pump should run 24 h a day, provide start-stop control _only_ at pump location.

(7) Pumps drawing from tanks should be provided with low water cutoff switches.

(8) Except for recirculating pumps pulling through diatomite filters, all recirculating pumps should be provided with full line size quick-opening basket strainers in their suction piping.

c. Heaters.

(1) Heaters should be shell and tube heat exchangers able to provide sufficient heat to

the pool to prevent freezing in the winter.
- (2) In small sizes, usually all the water is passed through the heater with the required temperature rise.
- (3) At high flow rates (gpm), it is usually more economical to pass only part of the water through the heater at an increased temperature rise, then blend it back with the bypassed water.

d. Chlorination equipment.

- (1) Hypochlorinators should be small, adjustable, positive displacement pumps (chemical feeders), built specially for the service, drawing sodium hypochlorite from a container and injecting it into the filter discharge piping to the pool after the last (hot-water-heater) connection.
- (2) The hypochlorinator must be provided with a plastic container to hold the required sodium hypochlorite.

    - (a) Check that the units selected have a discharge pressure capability greater than that in the piping that they connect to.
    - (b) The table gives suggested <u>minimum</u>-sized hypochlorinators:

System volume, gal	Suggested size, gph[a]
10,000	0.14
18,000	0.25
42,000	0.58
60,000	0.83
72,000	1.0
114,000	1.6
180,000	2.5
360,000	5.
720,000	10.
972,000	13.5
1,300,000	18.

[a]Capacity at maximum adjustment rating.

(3) Hypochlorinators should be interlocked with the filter recirculating pump so that they operate only when the pump is running.

G. Gas systems.

1. Principles of design.

   a. Provide gas to kitchen equipment, all other equipment requiring same, laboratory outlets, and elsewhere as required by the program.
   b. Gas should normally be supplied from the utility company's street mains. Where no street mains are available, supply must be from a liquified petroleum (LP) "bottled" installation.
   c. Contact the local utility company for the availability of street (natural) gas in the area. Obtain drawings showing locations and sizes of gas mains in the area. Obtain pressure and Btu/ft^3 content of the available gas.
   d. Gas service is usually low pressure (1/2 psi or less); however, sometimes it may be medium or high pressure and require reduction.
   e. Design gas systems in accordance with the utility company's requirements, NFPA Standard 54, and the requirements of applicable codes.
   f. Investigate carefully the requirements of equipment requiring gas; calculate the branch piping as well as the mains and risers on the basis of the requirements of the equipment.
   g. Loads for equipment should be taken from the manufacturer's ratings.
   h. Use factor for commercial kitchen equipment should be 100 percent. Simultaneous use factors for dwelling units (65,000 Btu/h per dwelling unit) should be as follows:

Gas use factors for apartments (cooking only)	
No. of Apartment units	Percent use factor[a]
1	100
2	70
3	60
4	55
5	50
6	45
7	42
8	40
9	37
10	35
11	34
12	32
13	32
14	30
15	30
16	28
17	28
18	28
19	27
20	26
50	18
80	15
100	14
150	12
200	11
250	11
300	10
400	9
500	8
600	7

[a] From Figure 9-57b, Gas Engineers Handbook, published by the Industrial Press; used by permission of the American Gas Association.

i. Load for a laboratory outlet should be 5,000 Btu/h for small burners and 10,500 Btu/h for large burners. Find out from the owner which type he uses. Generally, schools use small burners and research and hospital laboratories use large burners. If in doubt, assume large burners.

j. Calculate laboratory gas demands using the following simultaneous use factors:

Number of outlets	Use factor, percent	Minimum
1 to 8	100	
9 to 16	90	9
17 to 29	80	15
30 to 79	60	24
80 to 162	50	48
163 to 325	40	82
326 to 742	35	131
743 to 1,570	30	260
1,571 to 2,900	25	472
2,901 - up	20	726

k. Branches serving one or two classrooms should have 100 percent use factor regardless of the number of outlets. Use factor for more than two classrooms may be 80 percent, and thereafter twice the normal laboratory use factor if this is less than 80 percent.

l. Simultaneous use factors must be used with judgement and modified to adapt to special conditions as they occur in the system.

m. Gas piping should pitch back to the meter wherever possible. All risers and trapped sections should be provided with drip pockets consisting of nipple and cap.

n. All risers and branches should be valved. Provide mains with sectionalizing valves at strategic locations. All connections for future extension should be valved.

o. In piping up gas meters, the inlet is always on the left and the outlet is on the right.

p. Wherever possible, gas mains and risers should be run exposed rather than concealed in shafts or hung ceilings. This is to prevent the possible accumulation of gas in these closed spaces due to even minute leakage from the gas piping system, which may explode if in the proper concentration and subject to an igniting spark of any kind. Where it is impossible to keep this gas piping out of hung ceilings, shafts and wall recesses, the HVAC project engineer is to be notified so that he can provide some sort of ventilation for these spaces to prevent the possible accumulation of gas.

q. No gas piping should be run in or through air plenum ceilings, clothes chutes, air ducts, dumb waiters, stair enclosures, or elevator shafts.

r. Gas risers in air shafts must be encased or separated by masonry construction.

s. Gas piping passing perpendicular through air plenum ceilings or ducts, without connnections, should be sleeved for the full length.

2. Pipe sizing.

a. Piping from the street main to the meter is usually sized by the utility company. We are responsible for sizing the piping from the meter to the point of use.

b. Piping after the meter should be sized on the basis of the cubic feet per hour loads and simultaneous use factors given above.

c. Demands with different use factors should be carried separately in calculations and only added together at the end.

d. Base pipe-size selection on a maximum friction loss of 0.3 in of water from the meter to the farthest horizontal point of use. Take a credit of 0.1 in of water gain per 15 ft of rise in sizing risers. Convert actual run of piping to equivalent developed length before calculating friction losses.

e. Size piping to include future anticipated loads and to provide for normal flexibility in laboratory areas.

f. Use the following sizes for laboratory bench piping:

	Small burners		Large burners	
Piping, in	Natural gas (5 cubic feet per hour)	Liquefied petroleum gas (2 cubic feet per hour)	Natural gas (10 cubic feet per hour)	Liquefied petroleum gas (4 cubic feet per hour)
3/8	1	2		1
1/2	2	3-6	2	2-3
3/4	3-6	7-18	3-4	4-10
1	7-16	19-40	5-8	11-21
1-1/4	17-30		9-16	22-37

1 duplex outlet = 2 single outlets

(1) Branches to equipment should be at least the size of the connection on the equipment.
(2) See Calculation Forms K, L, and M, Chapter 5, for sizing piping.

g. Use gas calculator or Table 4-26 to determine friction loss in piping.

Table 4-26 Data for sizing gas piping

Gas flow (cfh)	Nominal pipe size, in							
	1/2	3/4	1	1-1/4	1-1/2	2	3	4
	Pressure drop per 100 ft of pipe, inches w.c.							
5	0.01							
10	0.04							
15	0.08	0.01						
20	0.15	0.02						
25	0.23	0.03						
30	0.33	0.04	0.01					
40	0.59	0.07	0.02					
50	0.94	0.12	0.03	0.01				
75		0.25	0.06	0.02				
100		0.45	0.11	0.03	0.01			
125		0.69	0.17	0.05	0.02			
150		1.10	0.24	0.08	0.03			
175			0.33	0.11	0.04	0.01		
200			0.44	0.14	0.05	0.01		
250			0.68	0.21	0.08	0.02		
300			0.99	0.31	0.13	0.03		
350				0.42	0.17	0.04		
400				0.55	0.22	0.05		
500				0.87	0.34	0.08	0.01	
750					0.78	0.19	0.02	
1,000						0.34	0.04	0.01
1,250						0.53	0.07	0.02
1,500						0.79	0.09	0.02
2,000							0.17	0.04
3,000							0.40	0.09

Source: Reprinted, by permission, from Building Systems Design.

3. Piping.

   a. Low-pressure piping should be steel pipe with threaded malleable iron fittings; or in larger sizes, and for medium- or high-pressure piping, welded fittings.

H. Fire standpipe systems: combined fire standpipe and sprinkler systems (New York City).

   1. Principles of design.

      a. This section basically covers fire standpipe systems and New York City combined systems.

         (1) Refer to Section I for sprinkler systems and combined systems outside New York City.
         (2) Refer to Section I for further details on sprinklers.
         (3) Refer to Section C for further details on water services, tanks, etc.
         (4) Refer to Chapter 3, Section C.3 for further details on underground piping and water services.

      b. Provide fire standpipe systems and combined fire standpipe and sprinkler systems where required or allowed by the code and/or underwriters. Fire standpipe protection for buildings under construction should be referenced to the requirements of the code, the fire department and/or the underwriters.

      c. Design systems in accordance with the code, the requirements of the fire chief or fire marshal, and/or NFPA Standards 13 and 14. Except where only a few isolated sprinkler heads are involved, calculate sprinkler piping hydraulically using the densities dictated by the owner's insurance underwriters.

      d. All devices and equipment installed in the systems must be approved and listed by underwriters laboratories and/or factory mutual, and, in New York City also by the Board of Standards and Appeals.

      e. All projects involving street-pressure sprinkler systems, street pressure fire standpipe systems or fire pumps require hydrant tests on the mains in all streets that could be used to feed the building. Since these tests take time to get, they should be initiated as soon as possible at the start of a project. Have a hydrant flow test made by the local water department or water company, or the underwriters.

f. Buildings abutting one another (looking like one building) should have systems cross-connected with common siamese.

g. Unheated spaces should have automatic dry standpipe systems. Dry fire standpipe systems cannot be combined with sprinklers.

h. Systems should be supplied from at least one automatic source and siamese connections.

   (1) Check the code. Some codes require three- and four-way siamese instead of the normal two-way siamese.
   (2) In New York City, siamese connections are 3 in; elsewhere they are normally 2-1/2 in.

i. Where street pressure is sufficient (as determined from hydrant flow test), use it as the automatic source for the system or the lowest zone.

j. Where everyday static pressures in the system exceed the allowable limits, the system should be zoned to comply with the maximum pressures allowed by code and/or NFPA.

k. For systems or zones where street pressure is insufficient, the automatic source should be gravity tanks, automatic fire pumps, or hydropneumatic pressure tanks.

l. Where underground cross connections, yard fire mains, or fire pumps 1,000 gpm or larger are involved, provide a jockey pump.

m. Provide systems with two water services whenever possible; otherwise, consider using on-site storage with a single service. If a pond or lake is available, it may be acceptable to the underwriters as the required on-site storage. Where allowed by code, two domestic services can also serve as supply to fire pumps (New York City).

   (1) The system should have fixed connections from either the public (domestic) water supply or another water source such as a pond or lake,

but not both.

(2) Where the system is connected to the public water supply and a pond or lake exists on the site, which the fire department might use as suction for their pumpers pumping into a siamese, provide an approved back-flow preventer in the public water-supply connection.

n. Whenever practical, provide the 2-1/2-in fire department hose valves in the stair halls and the 1-1/2-in "First Aid" hose valves with racks of hose in the corridors adjacent to the stair halls.

o. Provide fixed orifice-type pressure reducers on 1-1/2-in valves where required, and adjustable orifice type on 2-1/2-in valves where required. The use of approved pressure-reducing hose valves on NFPA design systems increases the allowable zone height from 275 to 400 ft. In Boston, provide pressure reducers in all 2-1/2-in hose valves where fire pressure exceeds 78 psi.

p. In combination systems (in New York City, buildings over 100 ft high), sprinkler connections may be taken from the fire standpipe risers. Standpipe work (plumbing) will terminate in a control valve (with tamper switch) on the connection. Continuing from there will be sprinkler work. For alterations (adding sprinklers) in existing buildings, check with the local authorities about the minimum acceptable water supply for the combined system.

q. Sprinkler work should include water-flow alarm devices, tamper switches, piping, and sprinkler heads.

r. On floors where a pressure of over 175 psi could occur during a fire, provide approved pressure-reducing valves on the sprinkler connections.

s. Provide tamper switches on control valves when required by code, by the underwriters, or the owner, and on combined systems.

(1) In combined New York City systems, tamper switches are required on the control valves in the system connections from the tank.

(2) Also, water-flow alarm devices are required on main from the tank or tanks.

t. Risers should have the following minimum sizes:

	New York City	NFPA
4"	Maximum 150 ft	Maximum 100 ft
6"	Higher	Higher
8"	......	Express risers

u. Branches to hose valves:

Valve, in	New York City	NFPA
Single 2-1/2	2-1/2 in maximum 4 ft	
	3 in maximum 25 ft	Use New York City requirements
	4 in longer	
Multiple 2-1/2	4 in	
Single 1-1/2	2-1/2 in	2 in maximum 50 ft
		2-1/2 longer

v. Provide control valves on all risers and branches with three or more hose outlets.

  (1) In New York City, provide single riser systems with intermediate control valves every 100 ft.

  (2) In New York City, riser control valves not located in a stair enclosure should be motor-operated, controlled from either the fire pump room or the entrance floor.

w. In an NFPA design system involving a gravity tank, the risers should be cross-connected, both top and bottom (also valved top and bottom), except where a dead riser is run down to feed more than one building.

x. Provide roof manifolds when not in a heated space with a gate valve below and an autoball drip to prevent freezing. The gate valve must be operable from the manifold location (stem or enclosing pipe flashed through the roof).

y. Hose lengths should be as allowed by the code or underwriters, whichever is shorter, remembering that the allowable stream throw is in a straight line and can't go around a corner. New York City -- 125 ft.

- z. Nozzles should be adjustable fog type. In buildings such as telephone company buildings, use model without straight stream.
- aa. Provide an alarm panel for the alarms from gravity fire tanks (high and low), fire pumps (current failure and pump running), hydropneumatic tanks (high and low water, and high and low air pressure), water-flow alarm devices, alarm valves, dry valves, deluge valves, CO^2 systems, Halon systems, tamper switches, and others as required for the project. Provide terminals on the panel for central station connection. The alarm panel is usually provided by the sprinkler contractor for a combined system.
- bb. Where required by code or allowed by code and requested by owner, provide manual dry standpipe system. This system normally has only the siamese as a source of water supply. Some codes also require a checked 3/4-in connection from the domestic main to keep water in the system at all times to be sure that all hose valves are closed.

2. Piping.

   a. Underground piping, except as noted, should be cement-lined, cast-iron or ductile-iron bell and spigot water pipe and fittings.
   b. Underground piping from building to sidewalk type siamese should be red brass pipe with fittings as for inside building (New York City tall buildings).
   c. In New York City, piping inside building should be standard weight (extra heavy when below 657 ft from top of system) steel pipe with code rated malleable cast-iron or cast-steel fittings or Victaulic malleable iron fittings and couplings.
   d. Sprinkler piping should be standard-weight steel pipe with standard-weight cast-iron threaded or flanged fittings, or Victaulic malleable iron fittings and couplings.
   e. Elsewhere, unless otherwise required by local code, piping inside building should be standard-weight (extra heavy when system fire pressure exceeds 300 psi) steel pipe with malleable iron or cast-iron fittings as listed in following table.
   f. Valves should be IBBM or cast steel as per

following table.

(1) Fire pump and jockey pump discharge check valves should be spring type when the lift to the top of the system is 35 ft or more.
(2) Control valves 6 in and larger, except at fire pumps should be provided with bypass valves.

g. Pipe, fitting, and valve-pressure ratings.

Height from top of system, ft	System fire, psi	New York City[a]			Elsewhere[a]		
		Schedule pipe	psi wwp fittings	psi wwp valves	Schedule pipe	psi wwp fittings	psi wwp valves
0- 115 4" 6"	165 155	40	350 MCI	150 (175)	40	Std MI 300 Std CI 175	175
116- 270	230	40	350 MCI	250 (400)	40	Std MI 300 Exh CI 400	400
271- 425	300	40	350 MCI	350 (500)	40[b]	Exh MI 800 Exh CI 400	500
426- 657	415	40	500 MCI	Class 300 (720)	80	Exh MI 800 Class 300 steel 720	Class 300 (720)
658-1,112	625	80	800 steel	Class 400 (960)	80	Class 400 steel (960)	Class 400 (960)

[a] Unless otherwise required by code, 800 psi wwp Victaulic malleable iron fittings and couplings may be used.
[b] Use Schedule 80, over 300 psi.

h. Piping subject to alternate wetting and drying, such as drain piping, test piping, siamese connection between siamese and check valve, and dry systems shall be galvanized.
i. Except as noted, calculate friction in fire piping on the basis of C = 120. Use C = 100 for dry, deluge, and preaction systems.
j. Provide wet piping running through unheated areas or otherwise exposed to possible freezing with electric heating cable and insulation. Check with project underwriters; some (FM) require heating pipe tracer rather than electric cable.

3. Equipment.

a. Size fire pumps as required by the code, the fire

chief or fire marshal, or the underwriters, whichever is greater. Since sizing requirements in NFPA Standard 14 are excessive, try to get the underwriters to accept a more reasonable size (500 or 750 gpm).

   (1) For motor-driven pumps, select 3500 rpm rather than 1750 rpm whenever a choice is available.
   (2) Refer to NFPA Standard 20 for data on fire pumps.

b. Calculate fire-pump heads by adding the pressure required at the top of the system or the zone served (converted from psi to feet), the lift from the pump to the top hose valve, and the friction in the system from the pump to the top hose valve minus the minimum available suction pressure.

   (1) In calculating the friction in a one zone system, assume the full gpm of the pump passing through the cross connection and/or the site fire main and up to the top of the riser serving the roof manifold.
   (2) In calculating the friction in the lower zone of a multizone system, assume the full gpm of the pump passing through the cross connection, splitting up the risers, and through the intermediate cross connection.
   (3) In a two-zone system, figure the friction in the top zone as for a one-zone system.
   (4) Where there are more than two zones, figure the friction in the top zone for a one-zone system and the lower zones as for the system friction in the lower zone of a two-zone system.

c. In determining the casing pressure of the pump to be specified and the need for a relief valve and its setting, add the shutoff head on the pump (head at no flow, not the specified rated pump head) and the maximum anticipated suction pressure under no-flow conditions. This gives the maximum discharge pressure of the pump (under no-flow conditions).

d. Calculate the head for the jockey pump similarly to that of the fire pump, except that due to the small flow, the friction loss will be greatly reduced and possibly the minimum available suction

pressure increased.
e. When no emergency generators are being provided in the project, check with the underwriters if diesel engine drive is required for the fire pump.
f. Provide fire pumps with current failure and pump-running alarms.
g. Provide fire pumps with a valved cross-connection between discharge and suction (inside the control valves) for testing.

   (1) 3 in for 500 gpm pump.
   (2) 4 in for 750 gpm and larger pump.

h. Fire pump relief valves, when required, should be piped back into the suction, except where on-site storage tank is available.
i. Check the code and underwriters requirements for the fire pump room.
j. Refer to Sections C.3.b.(3) and C.3.b.(2) for data on square and rectangular gravity house tanks and tank-fill pumps. Refer to NFPA Standard 22 for data on round gravity tanks.
k. Refer to Section C.3.b.(4) for data on hydropneumatic pressure tank-fill pumps. Refer to Section O.5.c for data on air compressors for dry systems and hydropneumatic pressure tanks. Refer to NFPA Standard 22 for data on hydropneumatic pressure tanks.

   (1) Hydropneumatic pressure tanks should contain two thirds-water and one third-air.

      (a) For New York City designs, the quantity required is only the actual water, add 50 percent additional for the air to obtain the tank size.
      (b) For NFPA designs, the quantity stated is the tank size (both air and water).

   (2) Hydropneumatic pressure tanks are normally installed at the top of the system and maintain a minimum pressure of 75 psi between fires. If installed at the bottom of the system, the pressure must be 75 psi plus the lift to the top of the system, plus the friction in the piping up to the top of the system.
   (3) Provide hydropneumatic pressure tanks with

gauge glass, pressure gauge, and alarms. Provide fill and air lines with check and globe valves.

(4) For New York City designs:

(a) The fill pumps should be 65 gpm capacity delivered at the required tank pressure (normally 75 psi) with a 2-in fill pipe to the tank.
(b) The air compressor should have the capacity to raise the tank pressure from atmospheric pressure to the required tank pressure (normally 75 psi) in 3 h.
(c) The pump and air compressor should be provided with automatic controls to maintain the proper pressure and air-water ratio in the tank.

(5) For NFPA designs:

(a) The fill pump should have a minimum capacity to fill the tank in 4 h and be capable of delivering at the required pressure (normally 75 psi) with a 1-1/2-in fill pipe to the tank.
(b) The air compressor should have a free air capacity of 16 cfm for tanks 7,500 gal and smaller and 20 cfm for larger tanks; all at the required tank pressure (normally 75 psi) with a 1-in air pipe to the tank.

I. Sprinkler systems: combined sprinkler and fire standpipe systems.

1. Principles of design.

a. This section basically covers sprinkler systems and combined systems outside New York City.

(1) Refer to Section H for fire standpipe systems, and New York City combined systems.
(2) Refer to Section H for further details on fire standpipe systems and fire pumps.
(3) Refer to Section C for further details on water services, tanks, etc.
(4) Refer to Chapter 3, Section C.3. for further details on underground piping and water services.

b. Provide sprinkler systems and combined sprinkler and fire standpipe systems in accordance with the requirements of the code and/or underwriters. Fire standpipe protection for buildings under construction should be referenced to the requirements of the code, the fire department, and/or the underwriters.
c. Design systems in accordance with the code, the requirements of the fire chief or fire marshal, and/or NFPA Standards 13 and 14. Except where only a few isolated sprinkler heads are involved, calculate sprinkler piping hydraulically using the densities dictated by the owner's insurance underwriters.
d. All devices and equipment installed in the systems must be approved and listed by underwriters laboratories and/or factory mutual; and, in New York City, also by the Board of Standards and Appeals.
e. All projects involving street pressure sprinkler systems, street pressure fire standpipe systems, or fire pumps require hydrant tests on the mains in all streets that could be used to feed the building. Since these tests take time to get, they should be initiated as soon as possible at the start of a project. Have a hydrant flow test made by the local water department or water company or the underwriters.
f. Buildings abutting one another (looking like one building) should have systems cross-connected with common siamese. Supervise services with water-flow switches at points of entrance.
g. Except as noted, sprinklers should be wet pipe. Unheated spaces should have dry systems. Areas where water damage would be critical should have preaction systems or wet pipe systems with on-off heads.
h. Systems should be supplied from at least one automatic source and siamese connections.

   (1) Check the code; some codes require three- and four-way siamese instead of the normal two-way siamese.
   (2) In New York City, siamese connections are 3 in; elsewhere they are normally 2-1/2 in.

i. Where street pressure is sufficient (as determined from hydrant flow test), use it as the automatic source for the system or the lowest zone.

j. Where fire pressures in the system exceed allowable limits, consult underwriters, code, and fire department regarding the possibility of zoning.
k. For systems where street pressure is insufficient, the automatic source should be gravity tanks, automatic fire pumps, or hydropneumatic pressure tanks.
l. Where underground cross connections, yard fire mains, or fire pumps 1,000 gpm or larger are involved, provide a jockey pump.
m. Provide systems with two automatic sources, whenever possible; otherwise, consider on-site storage with a single service. If a pond or lake is available, it may be acceptable to the underwriters as the required on-site storage.

   (1) The system shall have fixed connections from either the public (domestic) water supply or another water source such as a pond or lake, but not both.
   (2) Where the system is connected to the public water supply and a pond or lake exists on the site, which the fire department might use as suction for their pumpers pumping into a siamese, provide an approved back flow preventer in the public water supply connection.

n. To provide the proper protection, consider the design of the sprinkler systems carefully, mainly:

   (1) An automatic water supply of adequate pressure, capacity, and reliability.
   (2) Definite maximum protection area per sprinkler.
   (3) Minimum interference to discharge pattern by beams, bracing girders, trusses, piping, lighting fixtures, and air-conditioning ducts.
   (4) The correct location of sprinklers (deflectors) with respect to ceilings, or beams to obtain suitable sensitivity.
   (5) Coordinate with architect's reflected ceiling plan.

o. Design sprinkler systems with alarm valves in the basement, centrally located, and floor control valves with waterflow switches at the connection to the riser on each floor.
p. Where nonfireproof cooling towers are provided, sprinklers usually must be provided.

q. On floors where a pressure of over 175 psi could occur during a fire, use approved pressure reducing valves on the sprinkler connections.
r. Provide tamper switches on control valves when required by the underwriters or requested by the owner, and on combined systems.
s. For combined systems, 2-1/2-in and 1-1/2-in hose valves may be connected to wet sprinkler piping. Whenever practical, provide 2-1/2-in fire department hose valves in the stair halls and 1-1/2-in "First Aid" hose valves with racks of hose in the corridors adjacent to the stair halls.

   (1) Risers for combined systems should be 6 in.
   (2) In combined systems, provide control valves on all risers and branches with three or more hose valves.

t. Provide fixed orifice-type pressure reducers on 1-1/2-in valves where required, and adjustable orifice type on 2-1/2-in valves where required. The use of approved pressure-reducing hose valves on NFPA design systems increases the allowable zone height from 275 to 400 ft.
u. When not in a heated space, provide roof manifolds with a gate valve below and an auto ball drip to prevent freezing. The gate valve must be operable from the manifold location (stem or enclosing pipe flashed through the roof).
v. Hose lengths should be as allowed by the code or underwriters, whichever is shorter, remembering that the allowable stream throw is in a straight line and can't go around a corner.
w. Nozzles should be adjustable fog type.
x. Provide an alarm panel for the alarms from gravity fire tanks (high and low), fire pumps (current failure and pump running), hydropneumatic tanks (high and low water, and high and low air pressure), water-flow alarm devices, alarm valves, dry valves, deluge valves, CO^2 systems, Halon systems, tamper switches and others as required for the project. Provide terminals on the panel for central station connection.

2. Piping.

   a. Underground piping, except as noted, should be cement-lined cast-iron or ductile-iron bell and

spigot water pipe and fittings.
b. Underground piping from building to sidewalk type siamese should be red brass pipe with fittings as for inside building.
c. Sprinkler piping should be standard-weight steel pipe with standard-weight cast-iron threaded or flanged fittings, or Victaulic malleable iron fittings and couplings. Where fire pressure may be over 175 psi, use extra heavy fittings.
d. Valves should be 175 psi, wwp, IBBM. Where fire pressure may be over 175 psi, use 250 WSP valves. Fire pump and jockey pump discharge check valves should be spring type when the lift to the top of the system is 35 ft or more.
e. Check Section H.2.f. for fittings and valves for high-rise buildings.
f. Piping subject to alternate wetting and drying, such as drain piping, test piping, siamese connection between siamese and check valve, and dry system should be galvanized.
g. Except as noted, calculate friction in fire piping on the basis of C = 120. Use C = 100 for dry, deluge, and preaction systems.
h. Provide wet piping running through unheated areas or otherwise exposed to possible freezing with electric heating cable and insulation. Check with project underwriters, some (FM) require heating pipe tracer rather than electric cable.
i. Sprinkler pipe sizing can be based on hydraulic calculations (usually giving reduction of pipe sizes over those in the tables) or, in New York City, on reduced pipe sizing tables given for higher pressures. In New York City, the sprinkler service to and including the valve on the outlet side of the meter must be included in the plumbing contract, and the sprinkler work starts and continues from that point. Outside New York City, the sprinkler service is usually part of the sprinkler work; however, each locality should be checked for local practice.

3. Equipment.

    a. Fire pumps should be sized as required by the code, the fire chief or fire marshal, or the underwriters, whichever requirement is greater. As sizing requirements in NFPA Standard 14 are excessive, try to get the underwriters to accept a more reasonable size (500 or 750 gpm).

    (1) For motor driven pumps, select 3500 rpm rather than 1750 rpm, whenever a choice is available.

    (2) Refer to NFPA Standard 20 for data on fire pumps.

b. Calculate fire pump heads by adding the pressure required at the top of the system or the zone served (converted from psi to feet), the lift from the pump to the top hose valve or sprinkler head, and the friction in the system from the pump to the top hose valve or sprinkler head minus the minimum available suction pressure.

    (1) In calculating the friction in the system, assume the full gpm of the pump passing through the pump discharge, the cross connection, and/or the site fire main and up to the top of the riser or risers involved; or if a combined system, the riser serving the roof manifold.

    (2) For multizone systems, see Section H.3.b.

c. In determining the casing pressure of the pump to be specified and the need for a relief and its setting, add the shutoff head on the pump (head at no flow, not the specified rated pump head) and the maximum anticipated suction pressure under no-flow conditions. This gives the maximum discharge pressure of the pump (under no-flow conditions).

d. Calculate the head for the jockey pump similarly to the fire pump, except that due to the small flow, the friction loss will be greatly reduced, and possibly the minimum available suction pressure increased.

e. When no emergency generators are being provided in the project, check with the underwriters if diesel engine drive is required for the fire pump.

f. Provide fire pumps with current failure and pump running alarms.

g. Provide fire pumps with a valved cross connection between discharge and suction (inside control valves) for testing.

    (1) 3 in. for 500 gpm pump.

    (2) 4 in. for 750 gpm and larger pumps.

h. Fire pump relief valves, when required, should be piped back into the suction, except where on-site storage tank is available.

i. Check the code and underwriters' requirements for the fire pump room.

j. Refer to Section H.3 for data on gravity house tanks, hydropneumatic tanks, tank fill pumps, and air compressors for dry systems and hydropneumatic pressure tanks.

J. Fire extinguishers.

1. Principles of design.

   a. Provide fire extinguishers as required by the code and/or underwriters. Refer to NFPA Standard 10, Portable Fire Extinguishers.

   b. Check if the owner prefers to provide his own extinguishers rather than pay the contractor a markup on a piece of portable equipment.

   c. Provide 2-1/2-gal air-operated water-type fire extinguishers at all fire hose stations and elsewhere as required.

   d. Provide CO_2-type fire extinguishers at all kitchen cooking equipment, in each laboratory, in all machinery rooms, in garages, truck docks, and elsewhere as required. Extinguishers should be of the following sizes (weight of chemical, not total weight):

Kitchens	5 lb
Laboratories	5 lb
Small machinery rooms	5 to 10 lb
Large or main machinery rooms	10 to 15 lb
Garage and truck dock	10 to 15 lb

   e. ABC dry chemical-type fire extinguishers may be substituted for CO_2 type if so desired by the owner; however, they leave a residue. Dry chemical type are far superior to CO_2 type in windy locations. Extinguishers should be of the following sizes (weight of chemical, not total weight):

Kitchens	5 lb
Small machinery rooms	5 to 10 lb
Large or main machinery rooms	10 to 20 lb
Garage and truck dock	10 to 20 lb

    f. 10- to 20-lb dry chemical-type fire extinguishers may be alternated with CO_2 type in garages and truck docks.

    g. Mount corridor fire extinguishers in cabinets, and those elsewhere in cabinets at architect's direction.

    h. Consideration should be given to the total weight of the extinguisher, based on the persons who will have to lift it (men vs. women). However, garage extinguishers should be kept on the heavy side to minimize stealing.

K. Carbon-dioxide fire-extinguishing systems.

  1. Principles of design.

    a. Provide CO_2 protection in accordance with the requirements of the code (fire department directive) and/or underwriters.

    b. Design CO_2 system in accordance with the rules and regulations of the code and NFPA Standard 12.

    c. Provision of CO_2 systems should be considered for kitchen cooking surfaces, hoods (except where a fire protection system is built into the hood by the manufacturer), and exhaust ducts, alcohol storage rooms, anesthesia storage rooms, and the underfloor areas of computer rooms.

    d. Local application systems, such as those for cooking surfaces, kitchen hoods, and ducts should, upon actuation by fixed temperature detectors located below the hood and in the ducts, stop exhaust fans, shut down electrical power and/or gas to cooking equipment, sound local alarm, and actuate sprinkler alarm panel. In addition, pressure trips should be provided to release holding devices which will cause weighted dampers to close and contain CO_2 in the duct work being protected.

    e. Total room flooding systems should be actuated by smoke detectors and provided with a time delay so that the personnel working within the area may have sufficient time to vacate. Upon actuation of the system, pressure switches should

(1) Stop all supply and exhaust fans
(2) Shut down power to the computer equipment
(3) Sound local alarm
(4) Actuate sprinkler alarm panel

f. Upon actuation of the system, pressure trips should

(1) Release dampers to isolate area and prevent escape of CO_2 gas
(2) Release holding devices for doors, windows, etc.

g. Design the system so that the amount of CO_2 for the system is at least sufficient for protection of the largest single hazard or group of hazards which are to be protected simultaneously.

h. Locate and arrange storage containers and accessories so that inspection, testing, recharging, and other maintenance is facilitated and interruption to protection is held to a minimum. Locate storage containers as near as possible to the hazard or hazards they protect; but do not put them where they will be exposed to a fire or explosion.

i. Design piping arrangement and sizing to reduce friction losses to a reasonable minimum and take care to avoid possible restrictions due to foreign matter or faulty fabrication. Design piping to deliver the required rate of application at each nozzle.

j. Discharge nozzles should be suitable for the use intended and listed or approved for discharge characteristics.

k. Each system should have at least one manual control to actuate the system in its normal fashion. Manual controls should be so located as to be conveniently and easily accessible at all times, including the time of the fire. All normal operating devices should be identified as to the hazard they protect.

l. The manufacturer should be consulted for recommendations in the system design, piping arrangement, and sizing. The system should be approved by the manufacturer of the base equipment and installed in strict conformity with the manufacturer's instructions.

m. Where CO_2 systems are provided for concealed range-hood duct work, access doors should be provided adjacent to the location of the system actuators.

n. For systems requiring over 2,000 lb of CO_2, consider use of a low-pressure installation instead of a high-pressure cylinder system.

L. Halon fire-extinguishing systems.

1. Principles of design.

a. Provide Halon protection in accordance with the requirements of the code (fire department directive), and/or underwriters.
b. Design Halon system in accordance with the rules and regulations of the code and NFPA Standard 12A.
c. Provide Halon systems for underfloor areas of computer rooms, tape storage rooms, rare-book storage rooms, and art storage rooms, and computer rooms.
d. Actuate total room flooding systems by smoke detectors (cross zoned) and provide with a time delay so that the personnel working within the area may have sufficient time to vacate. Upon actuation of the system, pressure switches should

(1) Stop all supply and exhaust fans
(2) Shut down power to the computer equipment
(3) Sound local alarm (audible and visible, inside and outside)
(4) Actuate sprinkler alarm panel

e. Upon actuation of the system, pressure trips should

(1) Release dampers to isolate area and prevent escape of the Halon gas
(2) Release holding devices for doors, windows, etc

f. Design the system so that the amount of Halon gas for the system is at least sufficient for protection of the largest single hazard or group of hazards which are to be protected simultaneously.
g. Locate and arrange storage containers and accessories so that inspection, testing, recharging, and other maintenance is facilitated and interruption to protection is held to a minimum. Locate storage containers as near as possible to the hazard or hazards they protect, but not where they will be exposed to a fire or explosion.
h. Design piping arrangement and sizing so as to

reduce friction losses to a reasonable minimum and take care to avoid possible restrictions due to foreign matter or faulty fabrication. Design piping so as to deliver the required rate of application at each nozzle.

i. Discharge nozzles should be suitable for the use intended and listed or approved for discharge characteristics.

j. Each system should have at least one manual control to actuate the system in its normal fashion. Manual controls should be so located as to be conveniently and easily accessible at all times, including the time of the fire. All normal operating devices should be identified as to the hazard they protect.

k. Consult the manufacturer for recommendations in the system design, piping arrangement, and sizing. The system should be approved by the manufacturer of the base equipment and installed in strict conformity with the manufacturer's instructions.

M. Dry-chemical fire-protection systems.

1. Principles of design.

    a. Dry-chemical fire-protection systems should comply with all applicable local codes, all requirements of local authorities having jurisdiction, and NFPA Standard 17.
    b. Refer to NFPA Standard 17 for complete design criteria for these systems.
    c. Since these systems are specialized to the project requirements, consult the manufacturer for detailed design.

N. Foam fire-protection systems.

1. Principles of design.

    a. Foam fire-protection systems should comply with all applicable local codes, all requirements of local authorities having jurisdiction, and NFPA Standard 11.
    b. Refer to NFPA Standard 11 for complete design criteria for these systems.
    c. Since these systems are specialized to the project requirements, consult the manufacturer for detailed design.

O. Compressed-air systems.

1. Principles of design.

   a. There are four types of compressed-air systems, namely laboratory systems, hospital clinical (air used for patient treatment) systems, dental systems, and general building systems. Each must be a completely separate system.

      (1) Laboratory systems

         (a) Provide outlets, connected to a central system, in laboratories as required by the program.
         (b) The system should have rotary liquid-ring-type compressors with receiver, filters, and constant-pressure valve, delivering relatively dry, clean, oil-free air at 55 psi, and with a maximum 5 psi pressure drop in the piping, 50 psi at the outlets.
         (c) Provide lower-pressure requirements by pressure-reducing valves. Provide isolated higher-pressure requirements by isolated and/or subcentral compressor units of the required pressure.
         (d) Where especially dry air is requested or where the air piping runs through cold spaces (as in outside shafts), add an air dryer.
         (e) Allow 1 cfm (free air) (scfm) for each laboratory outlet with simultaneous use factors as follows:

No. of outlets	Use factor, percent	Minimum cfm
1 - 2	100	
3 - 12	80	3
13 - 38	60	10
39 - 115	40	25
116 - 316	30	50
317 - 700	20	95
701 - 1,880	15	145
1,881 - 4,400	10	285
4,401 - 16,000	5	445
16,001 - 80,000	2	800

(f) The branches serving one or two classrooms should have 100 percent simultaneous use factor regardless of the number of outlets. The use factor for more than two classrooms may be 80 percent and thereafter twice the normal laboratory use factor if this is less than 80 percent.

(g) Simultaneous use factors must be used with judgement and modified to adapt to special conditions as they occur in the system.

(h) Demands with different simultaneous use factors should be carried separately in the calculations and added together only at the end.

(i) Piping should be sized to include future anticipated loads and to provide for normal flexibility in the laboratory areas.

(j) Minimum pipe sizes should be:

    3/8-in branch to single or duplex outlet (short run)
    1/2-in other branches
    1/2-in riser
    3/4-in main

(k) The outlets must be an appropriate type for their locations and use, as specified.

(2) Hospital clinical systems.

  (a) Provide outlets, connected to a central system, as required by the program, to serve patient treatment.

  (b) Design these systems in accordance with the applicable requirements of medical gas systems including conformance with NFPA Standard 56F.

  (c) The system should have rotary liquid ring-type compressors with receiver, dryers, filters and constant-pressure valve delivering dry, clean, oil-free air at 55 psi, and with a maximum 5-psi pressure drop in the piping, 50 psi at the outlets.

  (d) Inlet piping to each compressor should be run independently through the roof as required by NFPA Standard 56F.

  (e) Outlets are normally provided at least

in the following locations and with the indicated required flows and simultaneous use factors. Check the program for the exact number and locations of all outlets.

Location	Volume per outlet (cfm)	Simultaneous-use factor
Major Operating rooms (two outlets per operating room)	2	100
Minor Operating rooms	2	100
Emergency rooms[a]	2	100
Trauma rooms	2	100
Plaster room	1	50
Delivery rooms (two outlets per room)	2	100
Cystoscopy and Special Procedures	1	20
Recovery (one outlet per bed)	2	50
Intensive Care (ICU) rooms (one outlet per bed)	2	50
Coronary Care (CCU) rooms (one outlet per bed)	2	50
Patients rooms (bedside outlets)[b]	1	10
Nurseries (one per four bassinets)	1	20
Special Care Nurseries (one per incubator)	1	100
Examination and Treatment rooms	1	10
Surgical Preparation rooms	1	10
Blood Donor rooms	1	10
Anesthesia rooms	1	10
Cardiac and Heart Catheterization rooms	1	10
Inhalation Therapy	1	10
High-level Radioisotope rooms	1	10
Low-level Radiation rooms	1	10
X-ray rooms	1	10
Pharmacy	1	10
Sterile Supply	1	10
Autopsy	1	100

Number of outlets: one per room except as noted.

[a] All outlets in the emergency department (area) should have 100 percent simultaneous-use factor.

[b] Sometimes one outlet per bed and sometimes one outlet per two beds.

(f) Carry the first outlet on the far end of a section of piping at full volume and apply simultaneous use factors to additional outlets only.
(g) Simultaneous-use factors must be disregarded when they produce a volume less than that of a single outlet.
(h) Simultaneous-use factors must be used with judgement and modified to adapt to special conditions as they occur in the system.
(i) Demands with different simultaneous-use factors should be carried separately in the calculations and added together only at the end.
(j) Be generous in pipe sizing, particularly in the smaller sizes, because there are so many unknowns involved, and the use of

compressed air in hospitals is on the increase.
- (k) Arrange the risers so that patients' outlets on any floor of any wing are divided between at least two risers.
- (l) A major operating suite should be served by more than one riser.
- (m) Provide the branches feeding outlets grouped with other medical gas outlets with ball valves mounted in a common multiple-valve box.
- (n) The minimum pipe sizes should be:

    3/8-in drops feeding a single outlet
    1/2-in other branches
    3/4-in riser
    3/4-in main

- (o) For outlets occurring in conjunction with medical gas outlets, see Section S.
- (p) Both central and local alarms should be provided as required for medical gases (see Section S).

(3) Dental systems.

- (a) Provide outlets in dental-chair pedestals connected to a central system, as required by the program for operating dental drills.
- (b) Design these systems in accordance with the applicable requirements of medical gas systems (Section S), including conformance with NFPA Standard 56F.
- (c) The system should have rotary liquid ring or other oilless-type compressors with receiver, dryer, filters, and constant-pressure valve; delivering clean, dry, oil-free air at 85 to 100 psi, with a maximum 5 psi pressure drop in the piping.
- (d) Contact the architect, owner, and/or tool manufacturer about the type of tools being considered and their volume and pressure requirements.
- (e) The minimum pipe sizes should be:

    3/8-in branch to a single outlet
     (short run)
   1/2-in other branches
   3/4-in riser
   3/4-in main

 (4) General building systems.

  (a) All other compressed-air requirements.
  (b) Provide compressed air of required quality in the required amounts and at the required pressures to satisfy the building's needs.
  (c) Check with the architect, owner, and/or equipment manufacturer about volumes, simultaneous-use factors, and pressures required.

2. General.

 a. Data in this section is based on 14.7-psia atmospheric pressure. In locations significantly different, cfm's, friction losses, and equipment selections must be adjusted accordingly.

 b. Make ample provision in the central plant and in the piping for anticipated future building expansion.

 c. Locate central compressed-air plants below the areas they serve with up-feed risers.

 d. Compressed-air piping should pitch back to the central plant wherever possible. Provide drains at the low points and trapped sections of the mains. The drains should be manual petcocks or automatic drain traps depending on the extent of the system served.

 e. All risers and branches should be valved. Provide mains with sectionalizing valves at strategic locations. All connections for future extension should be valved.

3. Pipe sizing.

 a. Size piping on the basis of the cfm loads and simultaneous-use factors given above.

 b. It should be noted that all cfm referred to in this section are free air at atmospheric pressure and are not the actual volumes of the compressed air in the piping which will be less due to compression and will vary depending on the pressure in the piping.

c. Base pipe-size selection in each case on the more stringent of the following requirements:

  (1) Maximum friction loss rate of 1 psi per 100 ft
  (2) Maximum friction loss to the farthest outlet of 5 psi
  (3) Maximum velocity of 4,000 ft/min

d. Table 4-27 gives the friction losses in 50 psi compressed-air piping. The actual run of the piping must be converted to equivalent developed length by adding fitting and valve allowances before using the tables. The maximum capacity not to exceed 4,000 fpm: 4 in - 1,600 cfm

Table 4-27 Pressure loss (psi) per 100 ft in 50 psi compressed-air piping

cfm of free air	Normal pipe size, in								
	1/2	3/4	1	1-1/4	1-1/2	2	2-1/2	3	4
5	0.30	0.03	0.01						
10	1.15	0.18	0.05	0.01					
15		0.40	0.11	0.03					
20		0.69	0.20	0.05	0.02				
25		1.14	0.31	0.07	0.03				
30			0.44	0.10	0.05				
35			0.61	0.14	0.06				
40			0.80	0.18	0.08				
45			1.00	0.23	0.10	0.03			
50				0.29	0.13	0.04			
60				0.42	0.18	0.05			
70				0.56	0.25	0.07	0.03		
80				0.74	0.33	0.09	0.03		
90				0.93	0.41	0.11	0.04		
100				1.15	0.51	0.14	0.05		
110					0.62	0.17	0.06		
120					0.73	0.20	0.08		
130					0.86	0.23	0.09	0.03	
140					1.00	0.27	0.11	0.03	
150						0.31	0.12	0.04	
175						0.42	0.16	0.05	
200						0.54	0.21	0.07	
225						0.69	0.29	0.08	
250						0.85	0.33	0.10	
275						1.03	0.40	0.13	
300							0.48	0.15	
325							0.56	0.18	
350							0.65	0.20	
375							0.74	0.23	
400							0.84	0.27	
450							1.06	0.32	
500								0.42	
550								0.50	0.12
600								0.60	0.14
650								0.70	0.17
700								0.82	0.19
750								0.94	0.22
800								1.06	0.25
850									0.28
900									0.32
950									0.36
1,000									0.39
1,100									0.48
1,200									0.57
1,300									0.67
1,400									0.77
1,500									0.89
1,600									1.00

e. For friction losses in other than 50 psi systems, use Table 4-28 or consult standard friction tables for the pressures involved.

f. See Calculation Form N, Chapter 5, for sizing piping.

4. Piping.

   a. Piping carrying air for patient treatment, regardless of size, should be type L copper tubing with wrought copper or cast brass fittings with silver brazed joints, or red brass pipe with threaded cast brass fittings. All pipe, fittings, and valves should be specially washed for medical gas use and protected from contamination thereafter.

   b. Except as otherwise noted, piping 2 in and smaller should be type L copper tubing with cast brass or wrought copper solder joint fittings, or red brass pipe with threaded cast brass fittings.

   c. Except as otherwise noted, piping 2-1/2 in and larger can be galvanized steel pipe with galvanized threaded malleable iron fittings.

   d. For general building systems, small piping may be galvanized steel. In some industrial systems, piping may be welded steel.

   e. Buried piping must be adequately protected against frost, corrosion, and physical damage by installation within a pipe or conduit, with proper cover, and adequate corrosion protective coating.

   f. Control (shutoff) valves should be ball valves.

   g. Check valves should normally be spring-loaded type.

5. Equipment.

   a. Size air compressors to carry the peak anticipated load with one compressor out of operation.

      (1) Where future loads are involved, if the proportions are workable, a three compressor installation may be the more practical with the third compressor installed later.

         (a) This will require a smaller standby unit.
         (b) Also a three compressor installation will provide a smaller unit for operation during low-demand periods.

   b. Compressors for laboratory, hospital clinical and

dental systems should be rotary liquid ring type delivering nonpulsating clean, oil-free, moisture-free air.

   (1) If pressures required are over 100 psi, oilless piston type will have to be used.

   (2) In large installations, where chilled water is always available, recirculated seal water systems can be considered to conserve water usage.

   (3) Use the following minimum sizes for the air compressor intake from the roof:

cfm free-air capacity	Size, in
50	2-1/2
70	3
110	3
210	4
400	5

   (4) Because inlet pressure (friction loss in the inlet piping), if over 1/2 psi, affects the manufacturer's published air-compressor rating tables, inlet-pressure conditions, where the inlet piping has to be run up through the roof, should be checked with the manufacturer before finalizing equipment selections.

   (5) Since these air compressors have minimum seal water pressure requirements, check this required pressure and connect to a building system of at least that pressure.

c. For general building systems, depending on the use of the air and the pressures required, compressors may be rotary liquid ring type, oilless piston type, or oil-lubricated piston type, either air-cooled or water-cooled depending on the size. They are usually provided with after coolers.

d. Air dryers: hospital clinical and dental systems, and elsewhere when absolutely dry air is required, should be provided with air dryers.

   (1) Normally these should be refrigerated dryers giving a pressure dew point of 35°F.

   (2) Where a lower dew point is required, provide

        a combination refrigerated and desiccant dryer.
- (3) Where dryers are provided, they should be complete with filters, and the separate pipeline filters omitted.
- (4) Consider small local units for isolated requirements.
- (5) In small sizes, compressor, dryer, and receiver are available as a package unit.

e. The size of the receiver (tank), inlet mufflers, inlet filters, and constant-pressure valve should be as recommended by the compressor manufacturer for the ultimate installation.

f. Filters on the discharge main from the central plant should be duplex absorption type, each sized for approximately one half to two thirds of the ultimate load. If proportions are workable, it may be desirable to use three filters with one future. Provide automatic drain traps on the filters.

Table 4-28 Air-pressure loss, psi in 100 ft of clean commercial steel pipe

CFM of Free Air	½ Inch			¾ Inch			1 Inch			1¼ Inches			1½ Inches		
	80 lb.	100 lb.	125 lb.	80 lb.	100 lb.	125 lb.	80 lb.	100 lb.	125 lb.	80 lb.	100 lb.	125 lb.	80 lb.	100 lb.	125 lb.
10	.46	.38	.31	.11	.09	.08	.04	.03	.02	.0086	.0071	.0058			
20	1.74	1.42	1.17	.41	.34	.28	.13	.10	.08	.032	.026	.021	.014	.012	.010
30	3.84	3.13	2.54	.90	.74	.60	.28	.23	.19	.068	.056	.046	.031	.026	.021
40	6.93	5.55	4.53	1.55	1.28	1.05	.46	.38	.31	.116	.096	.079	.053	.044	.036
50	10.7	8.65	7.01	2.42	2.00	1.62	.73	.60	.49	.18	.146	.120	.081	.067	.055
60				3.47	2.84	2.33	1.02	.84	.69	.25	.21	.17	.12	.095	.078
70				4.73	3.85	3.14	1.36	1.12	.92	.34	.28	.23	.16	.13	.10
80				6.14	5.01	4.08	1.76	1.44	1.18	.44	.36	.30	.20	.16	.14
90				7.75	6.40	5.17	2.23	1.85	1.49	.55	.45	.37	.25	.20	.17
100				9.62	7.80	6.33	2.69	2.21	1.81	.66	.55	.45	.30	.25	.20
125				15.5	12.4	9.8	4.18	3.41	2.79	1.03	.85	.69	.46	.38	.32
150				23.0	18.1	14.4	5.75	4.91	3.99	1.47	1.20	.99	.65	.54	.44
175							8.10	6.80	5.45	2.00	1.64	1.32	.90	.73	.60
200							10.9	8.79	7.11	2.58	2.12	1.73	1.15	.95	.78
250										4.05	3.30	2.67	1.82	1.48	1.20
300										5.78	4.71	3.83	2.55	2.10	1.72
350										7.90	6.45	5.15	3.53	2.86	2.35
400										10.3	8.30	6.74	4.53	3.70	3.03
450													5.80	4.65	3.80
500													7.12	5.79	4.71

CFM of Free Air	2 Inches			2½ Inches			3 Inches			4 Inches			5 Inches		
	80 lb.	100 lb.	125 lb.	80 lb.	100 lb.	125 lb.	80 lb.	100 lb.	125 lb.	80 lb.	100 lb.	125 lb.	80 lb.	100 lb.	125 lb.
50	.024	.020	.016	.010	.008	.007									
60	.033	.027	.022	.014	.011	.009									
70	.044	.036	.030	.018	.015	.012									
80	.056	.046	.038	.023	.019	.015									
90	.070	.058	.048	.028	.024	.020									
100	.084	.069	.057	.035	.029	.023	.012	.010	.008						
125	.130	.107	.088	.055	.043	.036	.018	.015	.012						
150	.19	.15	.12	.074	.061	.050	.025	.021	.017						
175	.25	.20	.17	.099	.081	.067	.034	.028	.022						
200	.31	.26	.21	.128	.105	.086	.043	.036	.029						
250	.49	.40	.33	.195	.160	.131	.065	.054	.044	.017	.014	.011			
300	.70	.57	.47	.27	.23	.19	.092	.075	.062	.024	.020	.016			
350	.94	.77	.64	.37	.31	.25	.124	.101	.083	.032	.026	.022			
400	1.20	.99	.81	.48	.40	.33	.160	.131	.108	.041	.034	.028			
450	1.55	1.27	1.05	.60	.50	.41	.20	.165	.135	.051	.042	.035			
500	1.91	1.56	1.29	.75	.62	.50	.25	.20	.17	.062	.051	.042			
600	2.75	2.23	1.83	1.08	.89	.73	.35	.29	.24	.089	.073	.060			
700	3.67	3.00	2.45	1.43	1.18	.97	.48	.39	.32	.119	.098	.081			
800	4.90	4.00	3.23	1.87	1.54	1.25	.61	.50	.41	.154	.126	.104			
900	6.20	5.05	4.10	2.35	1.95	1.57	.77	.63	.54	.193	.159	.131			
1000	7.62	6.20	5.04	2.89	2.37	1.94	.94	.78	.64	.23	.19	.16	.075	.062	.051
1200	11.4	9.05	7.45	4.21	3.45	2.78	1.37	1.12	.92	.34	.28	.23	.105	.088	.072
1500	18.3	14.5	11.7	6.62	5.39	4.38	2.15	1.73	1.43	.52	.43	.35	.162	.135	.110
2000				12.0	9.66	7.80	3.77	3.09	2.52	.91	.75	.61	.28	.24	.192
2500							6.00	4.85	4.00	1.43	1.16	.95	.44	.37	.30
3000							8.60	6.98	5.67	2.05	1.65	1.35	.63	.52	.43
3500							12.0	9.65	7.80	2.78	2.25	1.84	.85	.70	.58
4000							15.9	12.6	10.2	3.55	2.91	2.37	1.11	.91	.75
4500										4.60	3.70	3.00	1.40	1.16	.95
5000										5.68	4.55	3.70	1.73	1.43	1.17
6000										8.12	6.58	5.34	2.45	2.01	1.65
7000													3.40	2.79	2.28
8000													4.46	3.63	2.89
9000													5.65	4.60	3.76
10,000													6.95	5.65	4.59

CFM of Free Air	6 Inches		
	80 lb.	100 lb.	125 lb.
1500	.063	.052	.043
2000	.110	.091	.075
2500	.170	.140	.115
3000	.24	.20	.164
3500	.33	.27	.22
4000	.43	.35	.29
4500	.54	.44	.36
5000	.67	.55	.45
6000	.94	.77	.63
7000	1.30	1.06	.87
8000	1.70	1.39	1.13
9000	2.15	1.76	1.43
10000	2.61	2.14	1.76
11000	3.23	2.63	2.14
12000	3.87	3.13	2.55

Lengths Other Than 100 Feet

The friction loss in pipe lengths shorter than 100 feet may be calculated proportional to the length. That is: for 50 feet, ½ the tabular figure; for 25 feet, ¼ the tabular figure, etc.

In pipe runs of more than 100 feet, the proportional method may be used providing the resultant friction loss does not exceed 9 or 10 psi. If it is greater, a more accurate check may prevent undersizing of pipes. Use the following method, based on the Fanning Equations from which these tables were derived:

Calculate first a conversion figure "K":

$$K = \frac{F_{100}(P - \tfrac{1}{2} F_{100})}{100}$$

Then: $F_L = P - \sqrt{P^2 - 2KL}$

Where:
F_{100} = tabular friction loss figure for 100 ft.
F_L = friction loss for length "L"
L = length of pipe in feet
and P = upstream pressure in pipe, psia

Source: Reprinted with permission of Ingersoll Rand Co.

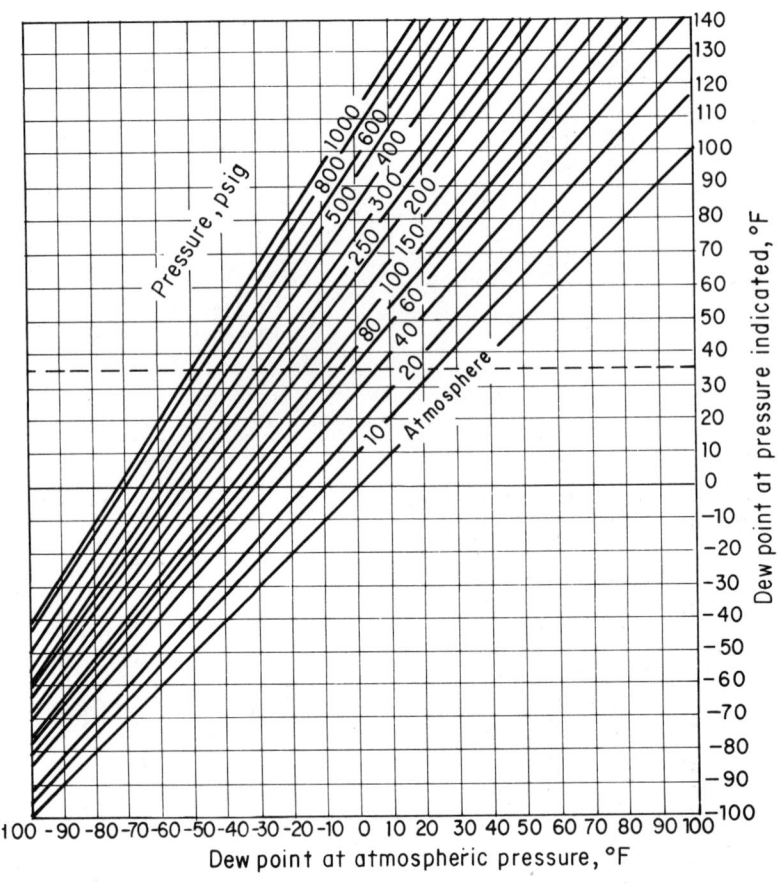

Dew point conversion:

To obtain the dew point temperature expected if the gas were expanded to a lower pressure proceed as follows:

1. Using "dew point at pressure", locate this temperature on scale at right hand side of chart
2. Read horizontally to intersection of curve corresponding to the operating pressure at which the gas was dried
3. From that point read vertically downward to curve corresponding to the expanded lower pressure
4. From that point read horizontally to scale on right hand side of chart to obtain dew point temperature at the expanded lower pressure
5. If dew point temperatures at atmospheric pressure are desired, after step 2 above read vertically downward to scale at bottom of chart which gives "Dew Point at Atmospheric Pressure"

Fig. 4-10. Dew point conversion chart. Hankison Corporation.

P. Vacuum air systems.

   1. Principles of design.

      a. Provide inlets connected to a central system in laboratories as required by the program.
      b. Provide inlets connected to a central system as required by the program in hospitals to serve patient treatment. Connect all inlets that serve patient treatment to a completely separate system from those serving the laboratory inlets.
      c. Make ample provision in the central plants and in the piping for anticipated future building expansion.
      d. Patient vacuum air inlets are normally provided at least in the locations shown on pages 4-168 and 4-169 and with the indicated required flows and simultaneous-use factors. Check the program for the exact number and locations of all inlets.

Location	cfm per inlet (at 15-in Hg)	Simultaneous-use factor
Operating room open heart, organ transplant, etc. (two per room)	3.5	100
Major Operating room (two per room)	2	100
Minor Operating room (two per room)	2	100
Cystoscopy and Special Procedures (two per room)	1	40
Emergency[a] (two per room)	2	100
Trauma room	1	100
Plaster (Fracture room)	1	100
Delivery room (two per room)	2	100
Recovery first inlet per bed	3	50
Recovery second inlet per bed	1	50
Recovery additional inlets per bed	1	10
ICU and CCU first inlet per bed	3	50
ICU and CCU second inlet per bed	1	50
ICU and CCU additional inlets per bed	1	10

Patient rooms surgical[b]
  Sometimes one inlet per bed
  Sometimes one inlet per two beds
	1	50

Patient rooms medical[b]
  Sometimes one inlet per bed
  Sometimes one inlet per two beds
	1	10

Labor rooms	1	20
Nurseries (one per four bassinets)	1	10
Special Care Nurseries (one per incubator)	1	40
Examination and Treatment rooms	1	10
Operating room bed holding areas	1	10
Surgical Preparation rooms	1	10
Blood Donor rooms	1	10
Anesthesia Work rooms	1	10
Cardiac and Heart Catheterization rooms	1	10
Deep Therapy rooms	1	10
Inhalation Therapy rooms	1	10
Electroencephalogram (EEG) rooms	1	10
Electrocardiogram (ECG) rooms	1	10
Electromyogram (EMG) rooms	1	10
Fluoroscopy rooms	1	10
High-level Radioisotope rooms	1	10
Low-level Radiation rooms	1	10
X-ray rooms	1	10
Bronchography room	1	100
Autopsy room	2	100
Central Supply room	1	10
Pharmacy	1	40

Number of inlets: one per room except as noted.

[a] All inlets in the emergency department (area) should have 100 percent simultaneous-use factor.

[b] Where patient rooms are interchangeable (surgical or medical), use 50 percent simultaneous-use factor for the first four rooms on the far end of the section of piping and 20 percent thereafter.

e. Allow 1 cfm at 15 in Hg for each laboratory inlet with simultaneous use factors as follows:

No. of inlets	Use factor	Minimum cfm
1 to 4	100	
6 to 12	80	5
13 to 33	60	10
34 to 80	50	21
81 to 150	40	40
151 to 315	35	61
316 to 565	30	111
566 to 1,000	25	171
1,001 to 2,175	20	251
2,176 to 4,670	15	436
4,671 up	10	701

The branches serving one or two classrooms must have 100 percent simultaneous-use factor regardless of the number of outlets. Use factor for more than two classrooms may be 80 percent, and thereafter twice the normal laboratory-use factor if this is less than 80 percent.

f. Simultaneous-use factors must be used with judgement and modified to adapt to special conditions as they occur in the system.
g. Demands with different use factors should be carried separately in the calculations and only added together at the end.
h. The central vacuum air plant should be located below the areas served with upfeed risers.
i. The vacuum pump discharge should be run up through the roof remote from all compressed air and other intakes, windows, etc.
j. The vacuum air piping should pitch back to the central plant wherever possible. Valved drain pockets and drain cocks should be provided at low points and trapped sections of the main.
k. Plugged cleanout connections should be strategically located throughout the system to allow a means for removing stoppages. These can be created by using a tee instead of an elbow or a cross instead of a tee.
l. The branches to risers should connect to the riser with a drop leg to facilitate the draining of the

contents of the piping back to the central equipment.
m. The branch piping to inlets should feed up to the inlets rather than down to the inlet wherever possible. Wherever downfeed branches are unavoidable, they should be connected to the top of the main branch to minimize the draining back to the inlet of the contents of the piping in the event of even momentary failure of the vacuum in the system.
n. All risers and branches should be valved. Mains should be provided with sectionalizing valves at strategic locations. All connections for future extension should be valved.
o. The branches feeding inlets grouped with other medical gas outlets should be provided with shutoff valves mounted in a common multiple-valve box. For branches 1-1/4 in and larger, valves may have to be in separate adjacent boxes.
p. The risers should be so arranged that patients' inlets on any floor of any wing are divided between at least two risers.
q. A major operating suite should be served by more than one riser.
r. The laboratory inlets should be the appropriate type for their location and use, as specified.
s. For inlets occurring in conjunction with medical gas outlets, see Section S.
t. Both central and local alarms should be provided for hospital clinical systems as required for medical gases (See Section S).
u. Data in this section is based on 14.7 psia atmospheric pressure. In locations significantly different, cfm's, friction losses, and equipment selections must be adjusted accordingly.

2. Pipe sizing.

   a. Size piping on the basis of the cfm loads and simultaneous-use factors given above.
   b. Note that the actual volume (cfm) of the air in the piping at any point is greater than the volume (cfm) of the room air (at atmospheric pressure) entering the system, due to expansion under vacuum (subatmospheric) conditions.

      (1) For 15 in Hg, the ratio is 2 to 1.
      (2) For 19 in Hg, the ratio is 2.7 to 1.
      (3) Be sure that the inlet requirements,

the friction-loss table, and the vacuum pump rating table all use the same cfm basis.

    (4) The data given herein is all compatible; all cfm at 15 to 19 in Hg.

c. Size piping to include future anticipated loads and to provide for normal flexibility in laboratory areas.

d. Base pipe-size selection in each case on the more stringent of the following requirements:

    (1) Maximum friction loss rate of 1 in Hg per 100 ft
    (2) Maximum friction loss to the farthest outlet of 4 in Hg
    (3) Maximum velocity of 5,000 ft

e. Table 4-29 gives the friction losses in the vacuum air piping: Convert the actual run of the piping to equivalent developed length by adding fitting and valve allowances before using the table.

f. The following is the maximum capacity not to exceed 5,000 fpm:

2 in	110 cfm
2-1/2 in	170 cfm
3 in	250 cfm
4 in	435 cfm
5 in	700 cfm
6 in	1,000 cfm

g. In large systems, consideration may be given to the fact that friction losses in piping 2 in and larger are less than average; therefore, some of this saving in friction can be used up in smaller sizes, still keeping the total under 4 in Hg. However, the loss in any pipe must be kept at 1 in Hg/100 ft or less.

h. Minimum pipe sizes should be:

    1/2-in branch to a single or duplex outlet (short run)
    3/4-in other branches
    3/4-in riser
    1-in main

i. See Calculation Form O, Chapter 5, for sizing piping.

Table 4-29 Pressure loss per 100 ft of pipe, in Hg

cfm[a]	\multicolumn{10}{c}{Nominal pipe size, in}
	3/4	1	1-1/4	1-1/2	2	2-1/2	3	4	5	6
1	0.02									
2	0.08	0.02								
3	0.16	0.04								
4	0.24	0.06	0.02							
5	0.36	0.08	0.02							
6	0.48	0.12	0.04							
7	0.60	0.16	0.04							
8	0.76	0.20	0.06							
9	0.92	0.24	0.08							
10	1.10	0.28	0.10	0.04						
15		0.60	0.18	0.08	0.02					
20		1.04	0.30	0.14	0.04					
25		1.12	0.47	0.22	0.06	0.02				
30			0.68	0.30	0.08	0.02				
35			0.90	0.40	0.10	0.04				
40			1.10	0.50	0.14	0.04	0.02			
45				0.62	0.16	0.06	0.02			
50				0.74	0.20	0.06	0.03			
60				1.00	0.26	0.10	0.04			
70				1.32	0.36	0.12	0.05	0.02		
80					0.50	0.16	0.06	0.02		
90					0.62	0.20	0.08	0.02		
100					0.76	0.24	0.10	0.03		
125					1.10	0.34	0.15	0.04		
150						0.55	0.23	0.06		
175						0.70	0.30	0.08		
200							0.39	0.08		
250							0.58	0.13		
300								0.20		
350								0.28		
400								0.35	0.10	
450								0.45	0.13	
500									0.17	
600									0.24	
700									0.30	0.10
800										0.12
900										0.16
1,000										0.20
1,100										0.24

[a] cfm is expanded cfm at 15 to 19 in Hg. NOTE: To use the NCG or Puritan-Bennett calculators, the cfm used must be cfm at atmospheric (room) pressure (scfm) (free air).

3. Piping.

   a. Piping 2 in and smaller should be type L copper tubing with wrought copper long turn solder joint fittings or red brass pipe with threaded cast brass long turn fittings.
   b. Piping 2-1/2 in and larger should be galvanized steel pipe with threaded galvanized cast-iron long turn drainage pattern fittings.
   c. Piping should be designed using Y or TY fittings instead of tees.
   d. Control (shut off) valves should be ball valves.
   e. Check valves should be soft-seated swing type.

4. Equipment.

   a. The vacuum pumps should be rotary liquid ring type.
   b. The vacuum pumps should be sized to carry the peak anticipated load at 20 in Hg with one pump out of operation. Where future loads are involved, if the proportions are workable, a three pump installation may be more practical with the third pump installed later.

      (1) This will require a smaller standby unit.
      (2) Also a three pump installation will provide a smaller unit for operation during low demand periods.

   c. Use the following minimum sizes for the vacuum pump discharge piping to the roof:

cfm capacity	Size, in
12	1-1/4
23	1-1/2
40	2
72	2-1/2
130	3
160	4
190	4
350	5
525	5

   d. Because back pressure (friction loss in the discharge piping), if significant, affects the manufacturer's published vacuum pump rating tables, back-pressure conditions on each project should be checked with the manufacturer before finalizing equipment selections.
   e. Vacuum air at the central plant should be maintained at 20 in Hg.
   f. The size of the vacuum tank should be as recommended by the pump manufacturer for the ultimate installation.
   g. In large installations, where chilled water is always available, recirculated seal water systems can be considered to conserve water usage.
   h. In systems serving patient treatment and in medical research laboratory systems, filters may be required in the discharge. Filters should be

in duplex. The filter pressure drop must be added to pipe friction losses in calculating discharge losses.

Q. Oral vacuum systems.

1. Principles of design.

    a. Connect oral vacuum inlets in dental-chair pedestals to a central oral vacuum system.
    b. Make ample provision in the equipment and piping for anticipated future building expansion.
    c. Calculate the system on the basis of 10 cfm inlet at 5-1/2 in Hg with normally a 100 percent use factor. A lower total system use factor should only be used if information from the owner on intended use justifies it. Vacuum producers and separator pumps should be duplex each full size.
    d. The central vacuum plant should be centrally located in the basement with upfeed risers.
    e. The piping should pitch back to the separator.
    f. The branch piping to inlets should feed up to the inlet rather than down to the inlet, wherever possible. Wherever downfeed branches are unavoidable, they should be connected to the top of the main or main branch to minimize the draining back to the outlet of the contents of the piping in the event of even momentary failure of the vacuum in the system.
    g. All risers and branches should be valved. All connections for future extension should be valved.
    h. The vacuum producer discharge should be connected to the chimney or flue, or run up through the roof.
    i. Plugged cleanout connections should be strategically located throughout the system to allow a means for removing stoppages. These can be created by using a TY instead of an elbow.
    j. Data in this section is based on 14.7 psia atmospheric pressure. In locations significantly different, cfm's, friction losses, and equipment selections must be adjusted accordingly.

2. Pipe sizing.

    a. Size piping on the basis of the cfm loads and normally a 100 percent use factor. Minimum pipe size should be 2 in.
    b. Note that the actual volume (cfm) of the air in

the piping at any point is greater than the volume (cfm) of the room (at atmospheric pressure) due to expansion under air vacuum (subatmospheric) conditions entering the system.

   (1) <u>Approximate expansion ratios</u>.

in Hg	Ratio
4	1.15
5	1.2
6	1.25
7	1.3
8	1.35

   (2) Be sure that the inlet requirements, the friction-loss table, and the vacuum producer rating table use the same cfm basis.
   (3) The data given herein is all compatible, all cfm (known as scfm) at atmospheric pressure.

   c. Figure 4-11 gives the friction losses in the oral vacuum piping.

   (1) Convert the actual run of piping to equivalent developed length before using the table.
   (2) Long mains may have to be oversized to cut down excessive friction losses.
   (3) The friction in the discharge piping should be kept minimal.

3. Piping.

   a. 2-in piping should be type L copper tubing with wrought copper long turn solder joint fittings or red brass pipe with threaded cast brass long turn fittings.
   b. Piping 2-1/2 in and larger should be galvanized steel pipe with threaded galvanized long turn cast iron drainage fittings.
   c. Piping should be designed using Y or TY fittings instead of tees.
   d. Control valves should be wafer or top-entry-type ball valves.

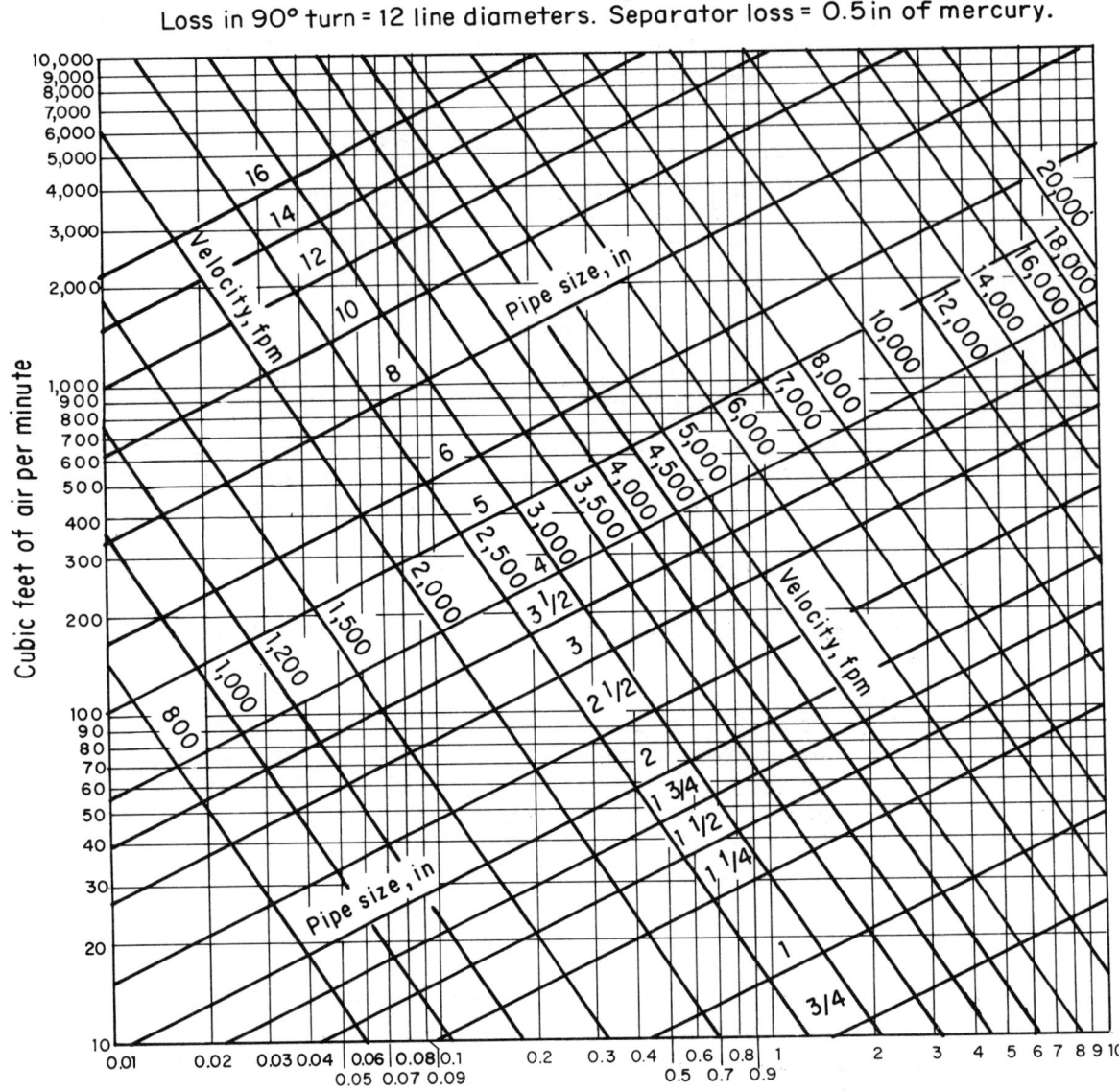

Fig. 4-11. Friction losses in oral vacuum piping. Compliments of the Spencer Turbine Co., Windsor, Conn.

4. Equipment

   a. Vacuum producers.

   (1) The vacuum producers should be duplex multi-stage centrifugal type with automatic control.
   (2) The vacuum head required for the system should be the sum of the following: inlet loss (5-1/2 in Hg), the system piping losses, the separator loss (0.5 in Hg) and the discharge loss.
   (3) Provide rubber flexible connections in both the suction and discharge piping at the vacuum producers.
   (4) Provide a check valve and blast gate in each discharge of duplex units, and a control valve in the suction.
   (5) Provide vacuum producers with surge control devices with silencer or use vacuum producers specially designed to overcome surges.
   (6) Where noise may be a factor, provide silencers on the discharge.

   b. Separator.

   (1) The separator should be a centrifugal hospital type with duplex drain pumps.

      (a) Pump capacity: 0.3 gpm per chair in use.
      (b) Pump control, automatic.
      (c) Provide a 6-in high curb around the separator with a floor drain capable of handling the pump-discharge rate.

R. Vacuum cleaning systems.

   1. Principles of design.

      a. Provide vacuum cleaning systems only when requested by the owner and/or architect. Their use is rare at the present time.
      b. Vacuum cleaning systems are not covered by code.
      c. Design the systems with the risers so located that all of the areas to be cleaned can be reached using not over 50 ft of 1-1/2-in hose. In hospitals, to minimize dangerous obstruction to corridor traffic (hose laying on the floor across the corridor), provide inlets on both sides of the corridor.

d. Assume an operator can clean 3,000 ft^2/h.
e. Make ample provision in the equipment and the piping for anticipated future building expansion.
f. In hospitals, use silent-type inlet valves in patient areas.
g. The hose and cleaning tools are usually provided by the plumber.
h. The central vacuum cleaning plant should be centrally located in the basement with upfeed risers.
i. The piping should pitch back to the separator.
j. The branch piping to inlets should feed up to the inlet rather than down to the inlet wherever possible.
k. The vacuum producer discharge should be connected to the chimney or flue, or run up through the roof.
l. Plugged cleanout connections should be strategically located throughout the system to allow a means for removing stoppages. These can be created by using a TY instead of an elbow.
m. Data in this section is based on 14.7 psia atmospheric pressure. In locations significantly different, cfm's, friction losses, and equipment selections must be adjusted accordingly.

2. Pipe sizing.

   a. Size piping on the number of simultaneous operators expected to use the system, not the number of inlets. Minimum pipe size should be 2 in.
   b. Each operating inlet uses 70 cfm. Use the following table for pipe sizing.

Size, in	Maximum number of operators
2	1
2-1/2	2
3	3
3-1/2	4
4	5
5	8
6	12
8	20

c. It should be noted that the actual volume (cfm) of air in the piping at any point is greater than the volume (cfm) of the room air (at atmospheric pressure) entering the system, due to expansion under vacuum (subatmospheric) conditions.

   (1) <u>Approximate expansion ratios</u>

in Hg	Ratio
2	1.1
3	1.1
4	1.15
5	1.2
6	1.25
7	1.3
8	1.35
9	1.4

   (2) Be sure that the inlet requirements, the friction loss table, and the vacuum producer rating table use the same cfm basis.
   (3) The data given herein is all compatible, all cfm (known as scfm) at atmospheric pressure.

d. Figure 4-12 gives the friction losses in the vacuum cleaning piping.

   (1) Convert the actual run of piping to equivalent developed length before using the above table.
   (2) Long mains may have to be oversized to cut down excessive friction losses.
   (3) The friction loss in the discharge piping should be kept minimal.

3. Piping.

   a. Normally piping should be black steel tubing and long turn fittings with brazed joints.
   b. In New York City, due to union rules, piping is standard-weight steel pipe with threaded long turn cast-iron drainage fittings.
   c. Design piping using Y or TY fittings stead of tees.

Hose loss = 1-1/2 times line loss. Loss in 90° turn = 12 line diameters. Separator loss = 0.5 in of mercury. Suction at tool = 2.0 in of mercury (average conditions)

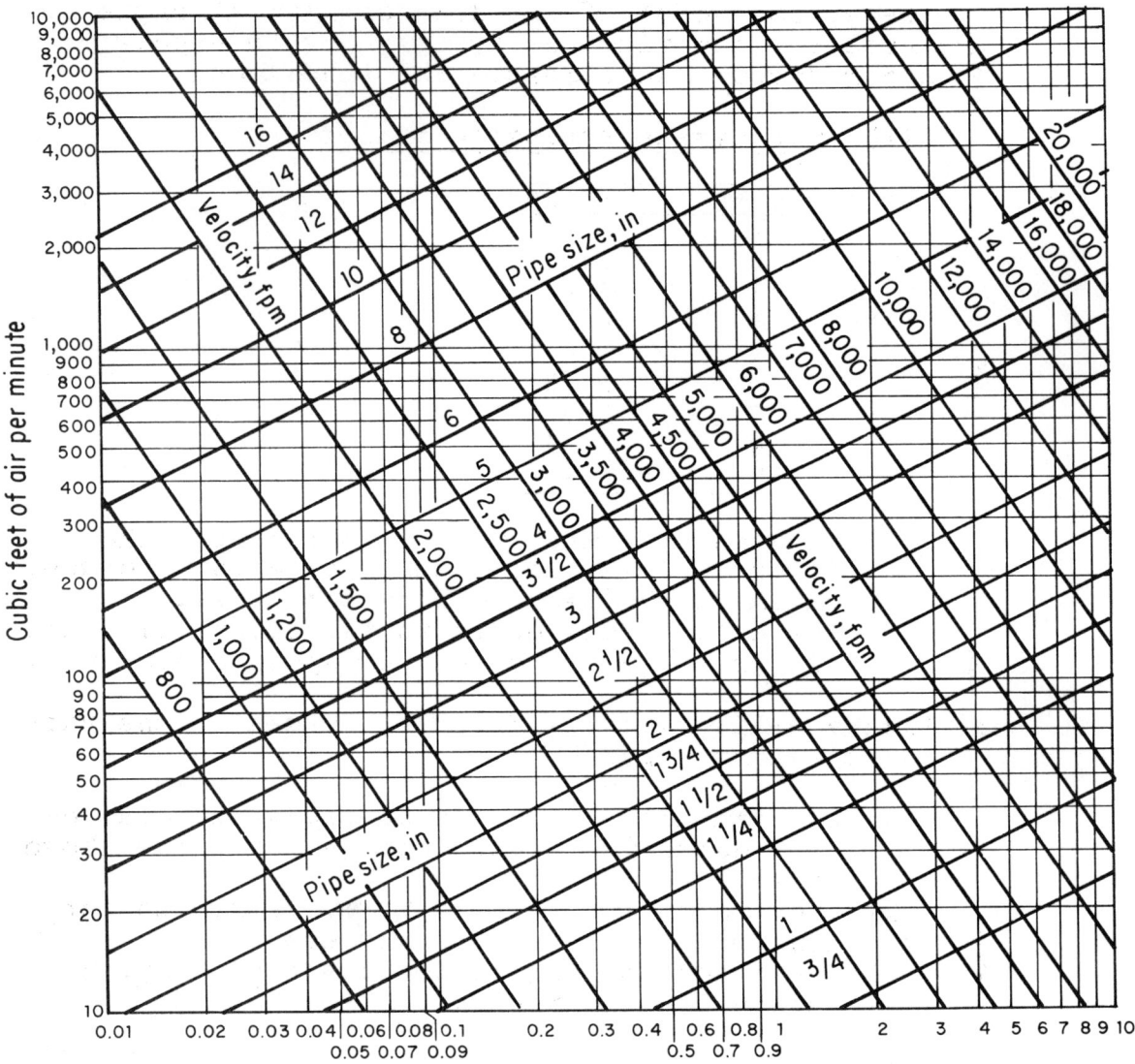

Fig. 4-12. Friction losses in vacuum cleaning piping. Compliments of the Spencer Turbine Co., Windsor, Conn.

4. Equipment.

   a. Vacuum producers.

   (1) The vacuum producer should be multistage centrifugal type. Control should be manual.
   (2) The vacuum head required for the system should be the sum of the following: the inlet loss (2 in Hg), the hose loss (1-1/2 times the loss in 1-1/2-in pipe), the system piping losses, the separator loss (0.5 in Hg for each separator), and the discharge loss.
   (3) Provide rubber flexible connections in both the suction and discharge piping of the vacuum producer.
   (4) Provide a check valve and blast gate in each discharge of duplex units.
   (5) Where required (check with manufacturer), provide surge control device with silencer.
   (6) Where noise may be a factor, provide silencer on the discharge.

   b. Separators.

   (1) Normally for systems of six operators or less, only a tubular bag-type separator is provided. For systems of eight operators or more, both a centrifugal-type separator (first in line) and a tubular bag-type separator should be provided.
   (2) For hospitals, provide a special hospital-type separator. For hospital-type separators, provide duplex pumps each with a capacity of 0.3 gpm per operator.
   (3) Since separators are high, carefully check available head room for installation.

S. Medical gas systems.

   1. General.

      a. Principles of design.

      (1) Design all medical gas systems in accordance with the requirements of NFPA Standards 56F (Non-Flammable Medical Gas Systems) and 50 (Bulk Oxygen Systems), local authorities, local codes, and for federally financed projects, HEW standards.

(2) Medical gas systems include the following systems: oxygen, nitrous oxide, medical compressed air, carbon dioxide, helium, nitrogen and mixtures of such gases when used for medical purposes. For medical compressed-air systems, see Section O.

(3) Make ample provision in the supply source and in the piping for anticipated future building expansion.

(4) Locate the risers with relation to the areas to be served in order to achieve minimum length branches.

(5) Arrange the risers so that patient's outlets on the floor of any wing are divided between at least two risers. A valved branch should preferably serve no more than 12 patient's outlets.

(6) A major operating suite should be served by more than one riser.

(7) Valves:

   (a) Control (shutoff) valves should be ball valves and installed in valve boxes with plastic windows. Where valves are grouped together, they should be installed in a common multiple-valve box. For branches 1-1/4 in and larger, valves may have to be in separate adjacent boxes.

   (b) Provide the main supply line with a shutoff valve so located as to be accessible in an emergency.

   (c) Provide each riser supplied from the main line with a shutoff valve adjacent to the riser connection.

   (d) Provide each lateral branch line serving patient's room outlets with a shutoff valve that controls the flow to the patient room outlets.

   (e) Locate a shutoff valve outside each anesthetizing location, operating room, delivery room, etc., in each line, so located as to be readily accessible at all times for use in an emergency. Arrange these valves so that shutting off the supply of gas to any one operating room or anesthetizing location will not affect the others.

(8) The type of outlet must be as approved or required by the owner. Where multiple outlets occur, they are usually mounted on a common plate in order from left to right: oxygen, nitrous oxide, compressed air, and vacuum.

(9) Medical laboratory facilities may require in isolated locations special gases such as oxygen, hydrogen, helium, carbon dioxide, nitrogen, or argon.

   (a) These requirements are usually supplied from local cylinders rather than from a central piped system.

   (b) Refer to the following NFPA Standards for special requirements for these gases: 50A, Gaseous Hydrogen Systems; 56C, Laboratories in Health Related Institutions.

b. Piping.

(1) All piping should be type L copper tubing with wrought copper or cast-brass fittings with silver brazed joints, or red brass pipe with threaded cast-brass fittings.

(2) All pipe, fittings, and valves should be specially washed for medical gas use and protected from contamination thereafter.

(3) Buried piping should be adequately protected against frost, corrosion, and physical damage by installation within a pipe or conduit, and with proper cover.

c. Alarm systems.

(1) Provide alarm systems as required in NFPA standard 56F.

(2) Locate alarm stations on the drawings.

(3) Include pressure gauges on local alarm stations.

2. Oxygen systems.

a. Principles of design.

(1) Large installations should be supplied from a bulk oxygen installation on the site. Small installations should be supplied from an oxygen cylinder manifold.

    1 to 100 outlets, manifold
    Over 100 outlets, bulk

(2) A rule of thumb for the preliminary sizing of the oxygen supply units (check with the supplier before final completion) is as follows:

   (a) Bulk systems: Allow 500 ft^3 per bed per month plus a reserve manifold of one day's supply.
   (b) Manifold systems: Allow one cylinder per bed for the main bank and double to provide for the reserve bank. 1 cylinder = 244 ft^3. Allow space in the manifold room for the storage of spare cylinders.

(3) Oxygen supply unit: the location of the unit (bulk or manifold) should be developed between the design team, the architect, and the supplier. Consideration should be given to truck access and minimum clearances from adjacent structures, as required by NFPA Standard 50.

(4) Outlets are normally provided at least in the locations shown on pages 4-185 and 4-186 and with the indicated required flows and simultaneous-use factors. Outlets may also be required in other rooms; check the program for the exact number and locations of all outlets.

Location	Simultaneous-use factor	Volume, Lpm
First Operating room (far end of a section of piping and all individual branches to operating rooms)	100	50 per operating room[a] (two outlets per operating room)
Second Operating room (on a section of piping)	100	30 per operating room[a] (two outlets per operating room)
Each additional Operating room (on a section of piping)	100	20 per operating room[a] (two outlets per operating room)
Emergency rooms[b]	100	Same as operating room
Trauma rooms	100	Same as operating room
Delivery rooms	100	Same as operating room
Cystoscopy and Special Procedures rooms	100	Same as operating room
Recovery rooms (one outlet per bed)		20 per outlet
one to eight outlets	100	
plus nine to twelve outlets	60	
plus thirteen to sixteen outlets	50	
plus additional outlets	45	
ICU rooms (two outlets per bed)	100	20 per outlet[c]
CCU rooms (two outlets per bed)	100	20 per outlet[c]
Other spaces such as:		20 per outlet[d]
Patient rooms (Medical and Surgical) (bedside outlets)		
Sometimes one outlet per bed		
Sometimes one outlet per two beds		
Labor rooms		
Sometimes one outlet per bed		
Sometimes one outlet per two beds		
Nurseries (check program for number of outlets)		
Special Care Nursery (one outlet per incubator)		
Examination and Treatment rooms		
Operating room bed holding areas		
Surgical Preparation rooms		
Blood Donor rooms		
Anesthesia Work rooms		
Plaster (Fracture) rooms		
Dental Surgery		
Cardiac and Heart Catheterization rooms		
Deep Therapy rooms		
Inhalation Therapy rooms		
Electroencephalogram (EEG) rooms		
Electrocardiogram (ECG) rooms		

(Continued on next page)

(Continued)

Location	Simultaneous-use factor	Volume, Lpm
Electromyogram (EMG) rooms		
Fluoroscopy rooms		
High-level Radioisotope rooms		
Low-level Radiation rooms		
X-ray rooms		
Endoscopy room		

a Where oxygen is used to power fluidically controlled anesthesia ventilators, increase Lpm volume by 40 percent.

b All outlets in the emergency department (area) should have 100 percent simultaneous-use factors.

c Where oxygen is used to power fluidically controlled ventilators, volume should be 40 Lpm.

d Simultaneous-use factors for other spaces: The first outlet on the end section of piping 20 Lpm, 100 percent use factor. For additional outlets on the section of piping 10 Lpm with the following use factors:

Outlets	Percent	Lpm	
1 - 3	100	Min	45
4 - 12	75	"	115
13 - 20	50	"	125
21 - 40	33	"	155
40 and over	25		

b. Pipe sizing.

   (1) Calculate pipe sizing on the Lpm loads and simultaneous-use factors given above.
   (2) Base pipe-size selection, in each case, on the more stringent of the following requirements:

      (a) Maximum friction loss rate of 1 psi per 100 ft.
      (b) Maximum friction loss to the farthest outlet of 5 psi.

   (3) Minimum size for risers should be 3/4 in. The minimum size for branches should be 1/2 in.
   (4) Outlet pressure should be a minimum of 50 psi.
   (5) See Calculation Form P, Chapter 5, for sizing piping.
   (6) Table 4-30 gives the friction losses in the oxygen piping. Convert the actual run of the piping to equivalent developed length by adding a fitting and valve allowance before using.

Table 4-30 Pressure loss per 100 ft of pipe, psi

Oxygen flow, Lpm	Nominal pipe size, in								
	1/2	3/4	1	1-1/4	1-1/2	2	2-1/2	3	4
50	0.04								
100	0.16								
125	0.25								
150	0.33	0.04							
175	0.48	0.06							
200	0.63	0.07							
250	0.99	0.11							
300	1.41	0.16	0.04						
400	2.51	0.29	0.07						
500	3.92	0.45	0.11						
750		1.02	0.24						
1,000		1.80	0.42	0.13	0.05				
1,250		2.81	0.66	0.21	0.09				
1,500			1.05	0.30	0.12				
2,000				0.67	0.22	0.05			
2,500				0.83	0.34	0.08			
3,000				1.19	0.49	0.11			
4,000				2.11	0.88	0.20	0.06		
5,000				3.30	1.36	0.32	0.10		
7,500					3.10	0.71	0.22	0.09	
10,000						1.27	0.40	0.16	
15,000						2.82	0.89	0.35	0.08
20,000						5.00	1.58	0.63	0.15
25,000							2.47	0.98	0.23
30,000							3.55	1.40	0.31
40,000								2.48	0.59
50,000								3.90	0.92

Source: Reprinted, by permission, from Building Systems Design.

3. Nitrous oxide systems.

   a. Principles of design.

      (1) A rule of thumb for the preliminary sizing of the inside nitrous oxide manifolds (check with the supplier before final completion): Allow one cylinder per operating, delivery, or equivalent room for the main bank and double to provide for the reserve bank. Allow space in the manifold room for the storage of spare cylinders.

      (2) Nitrous oxide supply unit: Outside manifolds should be avoided due to cold-weather operational problems.

         (a) The location of the unit should be developed between the design team and the architect.
         (b) Consideration should be given to truck access and minimum clearances.

      (3) Nitrous oxide outlets are normally provided in the following locations and with the indicated required flows. Outlets may also be required in other rooms; check the program for the exact number and locations of all outlets.

Location	Volume, Lpm
First Operating room (far end of piping and all individual branches to operating rooms)	30 per operating room
Second Operating room (on a section of piping)	20 per operating room
Each additional Operating room (on a section of piping)	15 per room
Delivery rooms	20 per room
Emergency rooms	20 per room
Trauma rooms	20 per room
Anesthesia Work room	15 per room
Plaster (Fracture) rooms	20 per room
Endoscopy room	15 per room
Dental surgery	15 per room

b. Pipe sizing.

   (1) Calculate pipe sizing on the Lpm loads given above with a simultaneous-use factor of 100 percent.
   (2) Base pipe-size selection, in each case, on the more stringent of the following requirements:

   (a) Maximum friction loss rate of 1 psi per 100 ft.
   (b) Maximum friction loss to the farthest outlet of 5 psi.

   (3) Minimum size for risers should be 3/4 in. Minimum size for branches should be 1/2 in.
   (4) Outlet pressure should be a minimum of 50 psi.
   (5) See Calculation Form Q, Chapter 5, for sizing piping.
   (6) Table 4-31 gives the friction losses in the nitrous oxide piping. Convert the actual run of the piping to equivalent developed length by adding a fitting and valve allowance before using.

Table 4-31 Pressure loss per 100 ft of pipe, psi

Nitrous oxide flow, Lpm	Nominal pipe size, in								
	1/2	3/4	1	1-1/4	1-1/2	2	2-1/2	3	4
50	0.04								
100	0.16								
125	0.25								
150	0.33	0.04							
175	0.48	0.06							
200	0.63	0.07							
250	0.99	0.11							
300	1.41	0.16	0.04						
400	2.51	0.29	0.07						
500	3.92	0.45	0.11						
750		1.02	0.24						
1,000		1.80	0.42	0.13	0.05				
1,250		2.81	0.66	0.21	0.09				
1,500			0.95	0.30	0.12				
2,000				0.67	0.22	0.05			
2,500				0.83	0.34	0.08			
3,000				1.19	0.49	0.11			
4,000				2.11	0.88	0.20	0.06		
5,000				3.30	1.36	0.32	0.10		
7,500					3.10	0.71	0.22	0.09	
10,000						1.27	0.40	0.16	
15,000						2.82	0.89	0.35	0.08
20,000						5.00	1.58	0.63	0.15
25,000							2.47	0.98	0.23
30,000							3.55	1.40	0.31
40,000								2.48	0.59
50,000								3.90	0.92

Source: Reprinted, by permission, from Building Systems Design.

4. Nitrogen systems.

   a. Principles of design.

      (1) A rule of thumb for the preliminary sizing of the nitrogen manifolds (check with the supplier before final completion): Allow 1-1/2 cylinders per operating room requiring nitrogen, and double to provide for the reserve bank. Allow space in the manifold room for the storage of spare cylinders.
      (2) Develop the location of the supply unit between the design team and the architect. Consideration should be given to truck access and minimum clearances.
      (3) Provide nitrogen outlets in operating rooms for driving surgical tools when required by the program.
      (4) Outlets may also be required in other rooms. Check the program for the exact number and location of all outlets.
      (5) Contact the architect, owner, and/or tool manufacturer about the type of tools being considered and their volume and pressure requirements.

   b. Pipe sizing.

      (1) Calculate pipe sizing on the required cfm per outlet with a simultaneous-use factor of 100 percent.
      (2) Base pipe-size selection, in each case, on the more stringent of the following requirements:

         (a) Maximum friction loss rate of 2 psi per 100 ft
         (b) Maximum friction loss to the farthest outlet of 10 psi

      (3) Table 4-32 gives the friction losses in the nitrogen piping. Convert the actual run of the piping to equivalent developed length by adding a fitting and valve allowance before using.
      (4) The minimum size for risers should be 3/4 in. The minimum size for branches should be 1/2 in.

Table 4-32 Pressure loss (psi) per 100 ft of 160 psi nitrogen piping

cfm	\multicolumn{6}{c}{Nominal pipe size, in}
	1/2	3/4	1	1-1/4	1-1/2	2
5	0.11	0.01				
10	0.43	0.07	0.02	0.01		
15	0.96	0.12	0.04	0.01		
20	1.70	0.26	0.07	0.02	0.01	
25	2.66	0.42	0.11	0.03	0.01	
30		0.59	0.17	0.04	0.02	
35		0.81	0.22	0.05	0.02	
40		1.06	0.29	0.07	0.03	
45		1.34	0.37	0.09	0.04	0.01
50		1.65	0.46	0.11	0.05	0.01
60		2.37	0.66	0.15	0.07	0.02
70			0.90	0.21	0.09	0.02
80			1.17	0.27	0.12	0.03
90			1.48	0.34	0.15	0.04
100			1.83	0.43	0.19	0.05
110			2.21	0.51	0.23	0.06
120				0.62	0.27	0.07
130				0.72	0.32	0.09
140				0.83	0.37	0.10
150				0.96	0.42	0.11

c. Outlets.

(1) Outlets should be in a control box containing control valve, pressure regulator, pressure gauges, and special outlet connection.

T. Central soap systems.

1. Principles of design.

   a. Central soap systems are usually provided only when requested by the owner and/or architect. (You can suggest them where suitable.) Where only individual soap dispensers are being used, check with the architect whether they are to be provided by the plumbing contractor or the general contractor.
   b. Provide batteries of three or more lavatories with central soap-dispensing systems.
   c. Provide other lavatories, except patients' private lavatories, with individual soap dispensers.
   d. Provide batteries of three or more gang shower heads with central soap-dispensing systems.
   e. Provide other shower heads with individual soap dispensers.
   f. Where batteries of lavatories are over one another on two or more floors, they should be fed from a tank on the uppermost floor with valved branches

serving each floor. Not more than four floors should be fed from one tank (maximum 40 ft from tank to lowest outlet).

g. Wherever possible, locate tanks in janitor's closets. When located in toilet rooms, place in men's toilet rooms in preference to women's toilets whenever possible. Tanks should be surface-mounted or recessed as per architect's preference.

h. Where stacked toilet rooms exceed four floors (the limit of a single tank), a master tank can be located in a machinery space directly above the local tanks and piped to manually fill the local tanks. Check with manufacturer about suggested sizing.

i. Run piping concealed, except for branches, to surface-mounted tanks and basin-mounted dispensing valves.

j. Put floor control valves, unless in janitor's closets, in or behind the wall with satin-finish stainless-steel access doors.

2. Piping.

   a. Exposed piping should be chromium-plated standard-weight brass pipe with threaded standard-weight cast brass fittings.
   b. Concealed piping should be standard-weight black steel pipe with threaded standard-weight malleable iron fittings.
   c. All piping should be 3/8 in.

3. Equipment.

   a. Local tanks.

      (1) Recessed or exposed satin finish stainless steel of sizes as follows:

          1 gal,   up to 4 outlets
          2 gals,   5 to 8 outlets
          2-1/2 gals, 5 to 12 outlets
          3 gals,   9 to 15 outlets
          5 gals,   13 to 25 outlets

      (2) 2-gal size also available in baked white enameled steel.

   b. Master tanks.

                (1) Round stainless steel in 25-, 50-, 100- and 250-gal sizes.

        4. Outlets.

            a. Lavatory outlets should be liquid lather type, either wall- or basin-mounted type as desired.

        5. Individual dispensers.

            a. Individual lavatory dispensers should be chromium-plated liquid lather as follows:

                (1) Basin-mounted type.
                (2) Surface-mounted type.
                (3) Recessed type.
                (4) Built as part of a shelf.

            b. Individual shower dispensers should be chromium-plated liquid type with vertical valve.
            c. Shower outlets should be liquid type with vertical valve.

U. Gasoline systems.

    1. Principles of design.

        a. Gasoline systems must comply with all applicable local codes, all requirements of local authorities having jurisdiction, and NFPA Standard 30.
        b. Refer to NFPA Standard 30 for complete design criteria for these systems.
        c. The maximum storage tank size allowed in New York City is 550 gal.
        d. Provide 18-in diameter (clear opening) manholes in tanks 60 in and larger in diameter.
        e. System must be designed in accordance with EPA requirements for vapor recovery. Consult with manufacturer regarding available devices and local practice.

V. Insulation.

    1. Principles of design.

        a. Insulate all cold-water piping (except underground portions and exposed chromium-plated piping at fixtures and equipment) with 1/2-in thick fiber-glass

sectional pipe covering.
b. Insulate all chilled-water and chilled-water-circulating piping with 1 in thick on circulated piping, and 1/2 in thick on deadleg branches, fiber-glass sectional pipe covering.
c. Insulate all hot-water, tempered-water, and hot-water-circulating piping (except exposed chromium-plated piping at fixtures and equipment) with fiber-glass sectional pipe covering.

   (1) Normally insulation thickness is 1/2 in. on deadleg branches and 1 in. on circulated piping.
   (2) As an energy conservation measure, insulation thickness on circulated piping should be increased to the following:

1/2 - 1-1/2 in	1 in thick
2 - 4 in	1-1/2 in thick
5 in and larger	2 in thick

d. Where a tempered water line is run from a local master thermostatic valve, the word "tempered" should be included in the specifications along with "hot water" to prevent any loopholes in the specifications.
e. Insulate all horizontal drainage piping between storm-water drains and leaders or house drains, exposed horizontal storm-water drainage piping, and drainage piping carrying chilled water between fixture and stack or house drain as for insulation on cold-water piping, using nested larger-diameter covering over hubs.
f. Insulate all drainage and water piping in walls and ceilings of auditoriums, conference rooms, board rooms, and other quiet-type rooms with 1-in thick fiber-glass sectional pipe covering.
g. Provide all drainage, water, and wet-fire-protection piping in areas subject to freezing with electric-heating cable. Insulate with 1-1/2-in thick fiber-glass sectional pipe covering, oversized to fit over pipe and cable. Cable will be provided under electrical section of work.
h. Insulate swimming-pool recirculating piping from the hot-water heater to the pool (except underground portions) with 1-in thick fiber-glass sectional pipe covering.
i. When a decorative pool or fountain is winterized,

insulate recirculating piping from the hot-water heater to the pool (except underground portions) with 1-in-thick fiber-glass sectional pipe covering and insulate all other recirculating piping (except underground portions) as for insulation on cold-water piping.

j. When an outside decorative pool or fountain is not winterized, insulate all pool drain piping (except underground portions) from pool to drain valve as for insulation on cold-water piping.

k. Insulate all fittings and valves on insulated piping with premolded or mitered segments of fiber-glass insulation at least the thickness of the pipe insulation.

l. Insulate domestic water-meter assemblies with 1-in-thick fiber-glass blanket.

m. Except where they are packaged units insulated by the manufacturer, insulate hot-water heaters, preheaters, and hot-water tanks with 1-1/2-in-thick calcium silicate or rigid fiber-glass.

n. Insulate shell and tube heat exchangers (instantaneous hot-water heaters) as for hot-water circulated mains.

o. Insulate metal, gas- or oil-fired, hot-water heater flues with 1-in-thick calcium silicate.

p. Insulate domestic hydropneumatic pressure tanks with cork plastic coating. For a more perfect installation, use 1-in-thick rigid fiber-glass.

q. Jackets on pipe insulation: fire retardant type (ASJ).

r. No cost reduction, omitting insulation on branch piping, should be approved without first warning the architect and/or owner in writing of the noise problem often resulting from the uninsulated piping.

s. Check ASHRAE Standard 90-75 for criteria on energy conservation.

W. Vibration isolation.

1. Principles of design.

    a. Unless otherwise noted, mount all rotating equipment (except in-line circulators and vertical shaft sump and ejector pumps) on vibration isolation to prevent transmission of vibration to the building structure.

(1) Check with a consultant for unusual conditions.
(2) Check with a consultant if you feel that the isolated location of the equipment would justify omission of vibration isolation.

b. Provide all floor-supported equipment with a minimum 4-in foundation (housekeeping pad).

(1) Foundation must be higher where required to line up equipment connections.
(2) The bottoms of spring held inertia blocks should be 2 in above foundation.

c. Select vibration isolators for the lowest operating speed of the equipment.
d. Following is a listing of the types of vibration isolation used for plumbing equipment.

(1) 3/4-hp and smaller horizontal pumps, air compressors, and vacuum pumps; all central chilled-water units; rubber-in-shear supported rails.
(2) 1 to 3-hp horizontal pumps, jockey pumps, air compressors and vacuum pumps; rubber-in-shear supported 6-in inertia blocks.
(3) 5-hp and larger horizontal pumps, jockey pumps, vertical booster pumps, air compressors, and vacuum pumps; spring-supported inertia blocks.

(a) Inertia block thicknesses:

```
 5 to  15 hp:  6 in
20 to  50 hp:  8 in
60 to 100 hp: 10 in
Over  100 hp: 12 in
```

e. Large (water-cooled) piston-type air compressors require a massive spring-supported inertia block and flexible connections in their connected piping. Check with a consultant regarding this equipment.
f. Do not mount fire pumps on vibration isolation.
g. Equipment mounted on vibration isolation and closely connected to an immovable mass should have horizontal flexible sections in these connections.
h. Flexible connections 1-1/2 in and smaller should be braided flexible bronze tubing or wire-reinforced

rubber hose with threaded brass ends. Flexible connections 2 in and larger should be braided flexible bronze tubing or wire-reinforced rubber hose with builtin flanges. Flexible connections should be the minimum lengths given in the table.

Size, in	Bronze Tubing, in	SS Tubing, in	Flanged Hose, in	Threaded Hose, in
3/8	9	8	..	..
1/2	10	9	..	..
3/4	11	10	..	9
1	12	10	..	10
1-1/4	13	12	..	13
1-1/2	15	12	..	15
2	16	14	12	18
2-1/2	18	16	15	..
3	20	17	18	..
4	24	18	24	..
5	30	22	30	..
6	36	24	36	..
8	42	..	36	..

The above lengths are for straight-line installation, and will absorb nominal deflections. For definite deflections and angle installations, check manufacturer's recommendations.

i. Where space conditions prevent the use of the lengths given in the table, and the surrounding spaces are not critical areas, single-arch flanged reinforced flexible rubber expansion joints may be substituted. Length 2 to 8 in. is 6 in. and 10 to 20 in. is 8 in.

j. Provide hangers and supports on piping connecting

to equipment mounted on vibration isolation with mound-shaped rubber-in-shear and spring vibration isolators for a horizontal distance of 50 ft from the equipment.

X. Plumbing fixtures.

  1. General.

    a. Plumbing fixtures must be provided at least in the numbers required by code.

      (1) In normal office buildings, as a result of investigations of the subject, recommend approximately 1-1/2 times the code minimum for really adequate water closet, urinal, and lavatory facilities. Figures should be based on the legal population of the space and not the actual population.

        (a) Obtain anticipated population split between male and female from the architect and/or the owner. In the absence of this information, use 50/50 or preferably 60/60 to allow for future changes.

        (b) Encourage the architect to design double-ended public toilets so that the changing population ratio can easily be accommodated by moving the partition between the men's and women's toilets.

      (2) Insurance companies tend to install even more fixtures with the toilet rooms on each floor designed to the actual population (working from the code minimum up). Consult owner.

      (3) Provide plumbing fixtures for the handicapped as required by code.

    b. Plumbing fixtures should be vitreous china, AR porcelain enameled iron, AR porcelain enameled steel, or stainless steel of latest design and type for their intended use.

    c. Wherever possible, fixtures should be hung on the wall.

    d. Trim above and exposed below all fixtures should be chromium-plated brass.

      (1) Satin chromium plate should be provided only

when requested.

 e. Select supply fixtures to conserve water wherever possible and practical.

  (1) Except for wall-mounted type, provide all supply fixtures with automatic flow-control devices.
  (2) If automatic flow control devices are desired for wall-mounted supply fixtures, they must be installed in the piping in the wall leading to the fixture and provided with an access door for required accessibility.

2. Water closets.

 a. Water closets should be vitreous china, except in correctional institutions.
 b. Except in residences and apartments, water closets should be elongated rim bowls with open front and open back seats, preferably black.
 c. Water closets in residences and apartments may be either elongated rim or round front bowls with open front white seats with covers.
 d. Water closets should normally be flushometer-operated siphon jet type.

  (1) Tank type should be used only for special reasons, for example: Architect's insistence (VIP toilets), to keep the booster pump small enough to avoid a suction tank (New York City), in an alteration to avoid or minimize increasing existing pipe sizes, where insufficient pressure exists for the flushometers.
  (2) Blowout type should be used only in correctional institutions, possibly in schools or other places where they would receive abuse in use.

 e. The following are the pressures required at the flushometers for the various types of bowls:

  (1) Floor-mounted siphon jet: 15 psi
  (2) Floor-mounted blowout: 20 psi
  (3) Wall-hung siphon jet: 20 psi
  (4) Wall-hung blowout: 25 psi

 f. The special low-down tank super-quiet water closets

require a minimum of 30 to 35 psi (depending on the make and model) for an adequate flush.

g. Provide bowls at the required height for the handicapped.

h. Provide water closets in patient's rooms with bedpan lugs or slots, and a bedpan cleanser.

   (1) Check with owner regarding type of bedpan cleanser, among the following

      (a) Swinging arm on flushometer tail pipe
      (b) Hand-held spray hose; cold water only
      (c) Hand-held spray hose; hot and cold water
      (d) Hand-held spray hose with hand wall control
      (e) Hand-held spray hose; pedal-operated

i. Supports.

   (1) Support wall-hung fixtures on combination chair carrier and drainage fitting.

      (a) Single water closets may be nonadjustable with or without auxiliary inlets for other adjacent fixtures. Select allowable carrier types to suit the wall construction.
      (b) For batteries, carriers should be adjustable type.

3. Urinals.

a. Urinals should be vitreous china, wall-hung flushometer type.

b. The siphon jet type should be used if the architect is willing to provide privacy shields between the urinals. The siphon jet type is a good compromise on water usage and maintenance between the blowout (high water use, low maintenance) and the washout (low water use, high maintenance). If the architect does not want to provide privacy shields, then use the blowout type, except in isolated areas where noise is a factor.

c. Flushometers should preferably have disc handle in front.

d. Provide blowout and siphon jet urinals with vacuum breakers. Washout type only if required by code.

e. Supports.

(1) On block walls, urinals can be supported on builtin backing plates.
   (2) On "dry wall" stud partitions, urinals must be supported on chair carriers specially designed for the purpose.

4. Lavatories.

   a. Lavatories are usually vitreous china.

   (1) Stainless-steel lavatories are available. They should be used only on request.
   (2) Plastic lavatories integral with the countertop are now available, and the architect may supply this type, with the plumber supplying only the trim.

   b. On tile walls, lavatories are usually slab type.
   c. On other wall surfaces, lavatories are usually provided with integral backs.
   d. Lavatories for counters can be mounted on, in, or under the counter. The architect's preference should be clearly established. When the supply fixture is mounted on the counter (not on the fixture itself), it must be the spread type with an extra long spout.
   e. Wheelchair lavatories or slab-type lavatories with 6-in instead of normal 2-in escutcheons should be provided for the handicapped.

   (1) Water supplies should have straight stops and be mounted flat against the wall to clear the user's knees.
   (2) In some installations, it may be necessary to insulate the hot-water piping, as the user may have no feeling for heat and so could unknowingly burn himself.

   f. Where individual soap dispensers or central soap valves are being provided, they should preferably be mounted in a punching in the lavatory.
   g. Provide special shampoo lavatories in beauty parlors and barber shops when requested.
   h. Provide lavatories in beauty parlors and barber shops with special hair traps instead of the regular P traps.
   i. Supply fixtures.

(1) Water conservation should be kept in mind and 1/2-gpm spray-head faucets used wherever possible.

   (a) Approval from the architect and/or owner should be sought for the use of 1/2-gpm spray heads instead of 2-gpm aerators on lavatory supply fixtures.
   (b) Where appropriate, approval from the architect and/or owner should be sought for providing 95° tempered hot water to public lavatories instead of the usual hot and cold.

(2) Normally, supply fixtures should be combination type, center-set rather than spread (to save needless waste of money), unless the architect insists on the spread type. In living spaces and VP private toilets, provide pop-up wastes; elsewhere, open strainer wastes.

(3) For hospital-use lavatories, provide blade handles, gooseneck spouts, and open strainer wastes.

   (a) Check with the owner about whether he wants any fixtures provided with knee or pedal control.
   (b) Check with owner regarding use of aerators, spray heads, or plain ends.

(4) Provide blade handles on supply fixtures on lavatories for handicapped.

(5) Where only 95° tempered water is being supplied to the lavatory, use a single faucet. Use half a combination supply fixture only on the architect's insistence, because it is needlessly expensive.

(6) Self-closing push-button supply fixtures should not be used unless requested or there is a definite need for using them, such as in a correctional institution or where small children are involved.

j. Supports.

(1) On block walls, lavatories with backs can be supported on builtin backing plates and slab type on builtin wall carriers.

(2) On dry wall stud partitions, lavatories must be supported on chair carriers specially designed for the purpose.

5. Sinks.

   a. Counter sinks are usually stainless steel.

      (1) These can be integral with the countertop and thus will be provided by the architect.
      (2) If not integral, these can be mounted on, in, or under the counter. The architect's preference should be clearly established.

   b. Hospital-use sinks, not in a counter, are usually vitreous china wall-hung type.
   c. Shop sinks are usually AR enameled iron wall-hung type.
   d. Kitchen and pantry sinks including supply fixtures, except domestic, are usually supplied by the kitchen equipment subcontractor.
   e. Wall-hung sinks can also be stainless steel.
   f. Wash sinks can be AR enameled iron, wall- or pedestal-mounted. However, they are usually circular-, semicircular-, or corner-type pedestal-mounted with either terrazzo, stainless-steel, or plastic bowls.

      (1) For circular units, carefully investigate if there is a vent and whether the supplies are dropping from the ceiling or coming up through the floor because different catalog numbers are involved. A stainless-steel shroud can be provided to cover piping dropping from the ceiling.
      (2) Where circular or 54-in-diameter semicircular units may often be used by only one or two people, sectional foot controls should be specified to conserve water.

   g. Provide sinks in Fracture or Plaster rooms and barium sinks with special traps instead of regular P traps. Check if any other sinks require special traps.
   h. Supply fixtures.

      (1) Hospital-use sinks should be provided with gooseneck spout and blade handles.

                (a) Check with the owner about whether he wants any fixtures provided with knee or pedal control.
                (b) Check with the owner regarding use of aerators, spray heads, or plain ends.

            (2) Other sinks generally require combination supply fixtures with swing spout.

        i. Supports.

            (1) Wall-hung fixtures should be supported on chair carriers.
            (2) Wall-hung fixtures on dry wall stud partitions must be supported on chair carriers specially designed for the purpose.

    6. Flushing-rim service (clinical) sinks.

        a. Clinical sinks should be vitreous china wall-hung blowout type.
        b. Supply fixtures should include flushometer and combination supply fixture with blade handles.
        c. Sometimes bedpan cleansers are also provided. See bedpan cleansers under "Water closets."
        d. Fixtures should be supported on chair carriers.
        e. Fixtures on dry wall stud partitions must be supported on chair carriers specially designed for the purpose.

    7. Slop (service) sinks.

        a. Slop sinks should be AR enameled iron with integral back, set on a trap standard.
        b. Combination supply fixture with VB and hose end, should be wall-mounted above the fixture.
        c. On block walls, backs should be anchored to the wall.
        d. On dry wall stud partitions, backs should be anchored to backing plates fastened to the studs. Check with the architect to provide studs in this area strong enough to support these backing plates.

    8. Mop sinks.

        a. Mop sinks can be either precast terrazzo (any size) or precast molded stone (only listed sizes).
        b. Supply fixture should be wall-mounted combination

4-204

with VB, hose end, and 4 ft of hose.
   c. Sometimes mop sinks are built in place by the architect, in which case the plumber provides only the supply fixture and the drain.

9. Laundry trays.

   a. Laundry trays (tubs) are normally vitreous china, precast molded stone, one or two compartment, but can also be AR enameled iron with integral back or set in or under a counter.
   b. Supply fixtures should be combination swing spout type.
   c. Laundry trays (unless counter mounted) should be supported on chair carriers. On dry wall stud partitions, carrier must be specially designed for the purpose.

10. Drinking fountains (water coolers).

    a. Provide drinking fountains for the handicapped as required by code.
    b. Drinking fountains today are usually stainless steel. They are also available in vitreous china.
    c. Drinking fountains should be projecting, counter, semirecessed, simulated semirecessed, recessed, or water coolers as per the architect's preference.
    d. Unless drinking fountains are connected to a central chilled water system, they should be provided with their own cooler unit.

       (1) A cooler unit built into the wall below with a wall grille can be provided with any drinking fountain.
       (2) Semirecessed, simulated semirecessed, and unit water coolers have cooler units built into them.

    e. Unit water coolers should be wall-hung type; however, floor type may be more practical for alterations. Support wall-hung units on chair carriers. On dry wall stud partitions, carrier must be type specially designed for the purpose.
    f. Drinking fountains for the handicapped should be set with rim approximately 36 in. above the floor (minimum clearance under the fountain 30 in.) and with at least an 18 in. projection. Control should be a lever handle.

g. Units are available with two projecting fountains (one set at regular height and one set for the handicapped) on one stainless-steel wall plate and grille for the cooler unit.
h. Where glass or carafe filling is desired, faucets can be provided on recessed and semirecessed fountains.
i. Where powdered coffee or tea making is desired, unit water coolers can be provided with integral hot-water dispenser.
j. As water coolers are not provided by the manufacturer with integral stop valves, line control valves must be indicated on the drawings on piping to water coolers.
k. Supports.

   (1) Recessed and semirecessed fountains should be supported on builtin metal mounting frames.
   (2) Projecting and simulated semirecessed fountains on block walls should be supported on metal backing plates.
   (3) Projecting and simulated semirecessed fountains on dry wall stud partitions should be supported on metal backing plates fastened to the studs. Check with the architect to provide studs in this area strong enough to support the load.
   (4) Handicapped fountains on dry wall stud partitions should be supported on chair carriers specially designed for the purpose or the metal mounting frames of their cooler units.

11. Bathtubs.

    a. Bathtubs should be AR enameled iron or AR enameled steel with a nonskid bottom.
    b. Tubs without showers should have a two-valve tub filler.
    c. Tubs with showers should normally have a nonscald pressure-balancing mixing valve with volume control and diverter spout.
    d. Where children are involved, consideration should be given to the use of a hand shower with VB on a slide bar.
    e. Hospital institutional (pedestal) baths should have nonscald pressure-balancing mixing valve with volume control, spout, transfer valve, and hand shower with VB.

f. Tubs are available with 16-in apron and 14-in depth to keep piping above the floor slab back into the pipe space.
g. Tubs are available with outlet connection either under the overflow or under the drain.

12. Showers.

    a. Shower stalls may be provided by the architect, either field-built (plumber provides the drain), or prefabricated of metal or plastic.
    b. A prefabricated shower receptor may be provided by the plumber with the walls field-built by the architect.
    c. Shower stalls may be prefabricated units with terrazzo, molded stone, or plastic receptors, and metal or plastic walls.
    d. The materials of the shower stall and who provides what should be carefully checked with the architect. The plumber normally provides the supply fixtures.
    e. Supply fixtures should normally be a nonscald pressure-balancing mixing valve and ball-joint head.

        (1) For gang showers, provide wall-mounted rigid heads.
        (2) Where children are involved and in showers for the handicapped, consider the use of a hand shower with VB on a slide bar.
        (3) Where shower controls are located outside the stall, a thermometer should also be provided.
        (4) Where a thermostatic shower control valve is requested, it should be the combined thermostatic and pressure-balancing type.
        (5) For school gang showers, provide a master thermostatic mixing valve in a locked cabinet (under the instructor's control) on the hot water in addition to the individual shower control valves.
        (6) In the interest of water conservation, shower heads should be as follows:

            (a) Showers for important people, for clubs, and for residences and apartments: 3-1/2 gpm type.
            (b) Elsewhere: 2-1/2 gpm type.

13. Emergency-use fixtures.

    a. Emergency showers, eye washers, and face washers should be provided as required by code and/or the owner's requirements.
    b. Emergency showers are usually located in the corridor outside the laboratories, with a pull chain anchored to the wall.
    c. When located in a laboratory, they usually have a pull chain with a ring.
    d. Floor drains are usually not provided for emergency showers.
    e. Eye washers and face washers are usually stainless steel or vitreous china and may be wall-hung, counter-mounted, or pedestal-mounted. Control can be either hand or foot.
    f. Hand-held eye washers, either counter- or wall-mounted, are also available.
    g. An emergency shower and an eye or face washer can be combined in one unit.

14. Bedpan washer: Sterilizers.

    a. Bedpan washer: Sterilizers are available as follows:

        (1) Floor-mounted, wall-mounted, and builtin
        (2) Automatic and manual
        (3) Washer (without steam), washer-sterilizer (with steam)

    b. These units require separate 2-in vapor vents through the roof.

15. Can washers.

    a. Can washers are usually supplied by the kitchen equipment subcontractor; however, they are sometimes provided by the plumber, especially when they are in a remote location from the kitchen. Sometimes they are merely a hot and cold hose bibb over a curbed area.
    b. Can washers are normally of two types: cabinet type, which automatically washes both inside and outside of the can (Dean); and dish type, which automatically washes only the inside of the can (Bestov and Air Void).

(1) The cabinet type requires 40 psi with 10-1/2 gpm flow (315 gph hot water) for single units and 21 gpm flow (630 gph hot water) for double units.
(2) The dish type requires 30 to 50 psi with 7-1/2 gpm flow (225 gph hot water).
(3) All units require vacuum breakers. Check, VB may come as part of the unit.
(4) The dish type requires adjacent hot and cold hose bibb for hand washing of the outside of the cans.

    c. Cabinet type requires 140°F hot water and dish type requires 170°F hot water, if not provided with steam.

    d. Floor drains at can washers should be provided with buckets.

16. Hydrotherapeutic equipment.

    a. A hydrotherapeutic area may include any or all of the following equipment.

(1) Arm bath
(2) Leg bath
(3) Full-body bath (Hubbard tank)
(4) Hydrotherapeutic pool

    b. This equipment is sometimes supplied by the architect but more usually by the plumber.

    c. The baths are stainless steel. Check the manufacturer's catalog for the accessories that should be specified with each, including the hoist for the Hubbard tank. Inform the architect of the trolley beam requirements of the hoist.

    d. Particular attention must be given to thermostatic controls for these baths. As thermostatic valves are spec-rated at 45 psi, they must be selected to give the required flows at the available pressure, then the selected size must be converted to flow rate at 45 psi for specifying.

    e. Hydrotherapeutic pools are actually small swimming pools. See Swimming Pool section for design requirements.

    f. In the following table are the flow ratings and quantities of hot water used. Catalog number refers to those of Ille Electric Corp.

Catalog number	Type bath	gpm hot water[a]	gph hot water 120°F
HM-500	Arm	10	40
HM-505	Arm	10	40
HM-600	Arm, leg, hip	20	110
HM-655	Arm, leg, hip	20	135
HM-675	Arm, leg, hip	20	160
HM-680	Arm, leg, hip	20	185
HM-685	Leg, hip	20	145
HM-690	Leg, hip	20	145
HM-695	Leg, hip	20	120
HM-1000	Arm, foot, leg	10	50
HM-1005	Arm, foot, knee	10	60
HM-1010	Foot, ankle	10	35
HM-801	Full body	40-50	660

[a] Approximate gpm flow requirements and rating (at 45 psi) of size thermostatic valve normally provided.

17. Laboratory trim.

   a. Laboratory trim (cocks, supply fixtures, and cup drains) are often provided by the architect with the laboratory equipment, but sometimes by the plumber.
   b. Cocks and supply fixtures are usually cast brass with choice of the following finishes.

      (1) Chromium plate
      (2) Satin finish chromium plate
      (3) Clear plastic over chromium plate or satin chromium plate
      (4) Plastic (color-coded)

   c. All plastic units are also available.
   d. Air, gas, and vacuum cocks are usually ground key cocks. Sometimes the owner may request the more expensive needle-valve type. These ground key cocks are tested to 100 psi and advertised for 30 psi WP by Manufacturer's Trade Association; however, the manufacturers assure us that they can be safely used at 50 psi. Above 50 psi compressed air, use needle valves.
   e. All water outlets with serrated tips (hose ends) must be provided with a builtin VB especially designed for the purpose.
   f. On benches, gas outlets are always double, vacuum outlets are usually double, and compressed-air outlets are usually single except where the architect makes them double for esthetic reasons. All outlets in hoods are usually single and provided

with remote control valves.
g. Distilled water, demineralized (deionized) water, and reverse osmosis water outlets are available in brass tin-lined or silver-plated inside, stainless steel, aluminum, and PVC. Choice should be at least equal to the piping material. These outlets should always have self-closing valves.
h. Cup drains are available in stainless steel, lead, glass, and plastic. Check with the architect about his preference if the plumber is supplying them.
i. Traps are always provided by the plumber. Normally they are the same material as the waste piping.
j. Laboratory sinks including waste outlets (tail pieces) are usually provided by the architect. Check with the architect to be sure that he is including the tail pieces.
k. If any laboratory sinks are to be provided by the plumber, check with the architect and/or owner about his preference between stainless steel, plastic, ceramic, and alberene stone.

18. Fixture trim.

a. Kitchen fixtures are usually provided with supply fixtures and waste outlets (tail pieces) by the kitchen equipment subcontractor.
b. Hospital equipment (sinks), where supplied by the architect, normally include waste outlets (tail pieces). Supply fixtures for these sinks may be supplied with the sinks or may be provided by the plumber (the latter is preferable). Check with the architect.
c. Check with the architect to be sure that he is including the tail pieces.
d. Traps for the above are always provided by the plumber.
e. For supply fixture requirements see section on sinks.
f. Stop valves must be provided for all the above equipment, when supply fixtures are not provided by the plumber.

19. Toilet accessories.

a. Toilet accessories are usually provided by the architect. However, in some communities (for example, New York City), union rules require that metal items be installed by the plumber. This

point should be checked out on each project.
b. The toilet accessories must be securely fastened to the wall in a tamperproof manner.
c. The toilet accessories shall be chromium plated or satin finish chromium plated brass (not die cast zinc) or stainless steel.
d. If providing the toilet accessories, review the selection with the architect.
e. For soap dispensers, refer to the section on Central Soap Systems.

CHAPTER 5

CALCULATION FORMS

FORM A

ELECTRICAL CONNECTION DATA						
JOB _____ CURRENT CHAR _____ V ___ PH ___ WIRE ___ CY.					SHEET ___ OF ___ DATE _____	
ISSUED TO _____ V ___ PH ___ WIRE ___ CY.					REVISED _____	
MOTORS ___ HP OR LESS: ___ VOLTS ___ PHASE				STARTERS:		
MOTORS ___ HP OR OVER: ___ VOLTS ___ PHASE				TYPE	TYPE OF DISCONNECT	TYPE OF STARTING
EQUIPT. ___ KW OR LESS: ___ VOLTS ___ PHASE						
EQUIPT. ___ KW OR OVER: ___ VOLTS ___ PHASE						
MOTORS ___ HP OR LESS: FULL INRUSH STARTING						
MOTORS ___ HP OR OVER: REDUCED INRUSH STARTING						
NO. OF OVERLOAD DEVICES FOR 3 PHASE MOTORS ☐ 2 ☐ 3						
SYSTEM NO.	EQUIPMENT AND LOCATION	HP OR WATTS	STARTER TYPE	INTERLOCK SYST. NO.	SMOKE DET. HEAD LOC.	ACTUATING & REMOTE CONT. DEVICES & LOCATIONS	REMARKS

FORM B

NO.	EQUIPMENT	LOCATION	C.W.				GAS			DRAIN		
			USE	PR	gpm	SIZE	PR	BTU HR	SIZE	USE	gpm	SIZE
	ALARMS											

PLUMBING CONNECTIONS REQ'D FOR H.V.A.C. EQUIPT.

JOB _____ ISSUED TO _____ DATE _____

RETURN TO PLBG. DESIGNER AS EARLY AS POSSIBLE ACCOMPANIED WITH MARKED UP PRINTS.

FORM C   Sheet 1

JOB				SHEET NO. ___ OF ___ SHEETS		
SYSTEM STORM WATER DRAINAGE (SHEET 1 OF 3)				BY ___ DATE ___		
				RATE OF RAINFALL = YEAR PERIOD		
			INTENSITY DURATION = i			
C.B.Nº	DISTANCE WATER TRAVELS	CHANGE IN GRADE ELEVATIONS Δh	GRADE SLOPE	INLET CONCENT. TIME (min)	i RAINFALL INTEN. (in/hr)	REMARKS

FORM C   Sheet 2

JOB			SHEET NO. ____ OF ____ SHEETS
SYSTEM	STORM WATER DRAINAGE		BY _____ DATE _____
	(SHEET 2 OF 3)		

$Q = A.C. \, i$

C.B. N°	⌗ PAVED	100% (C)* ADJ. PAVED	⌗ GRASS	35% (C)* ADJ. GRASS	TOTAL ADJ. ⌗	$\frac{A.C. \, ⌗}{43{,}560}$ = A.C. IN ACRES	i	$Q = A.C. \, i$ = CFS

*VARIABLE

FORM C   Sheet 3

JOB																											
SYSTEM: SYSTEM WATER DRAINAGE (SHEET 3 OF 3) — SHEET NO. ___ OF ___ SHEETS — BY ___ DATE ___																											
PIPE SIZE																											
SIZE																											
INLET EL. OF ENTER. MH/CB																											
DROP																											
SLOPE %																											
FT. OF RUN																											
TO MH OR CB N°																											
OUTLET ELEV.																											
TOTAL CFS																											
CFS. C.B.																											
C.B. OR MH N°																											

FORM D

JOB _____ DATE _____ BY _____

| \multicolumn{9}{c}{STORM WATER DRAINAGE SIZING} |
LDR. NO.	LDR ⌀	PITCH	SIZE	MAIN	PITCH	SIZE	REMARKS

5-7

FORM E

JOB _____ DATE _____ BY _____

STORM WATER SIZING SHEET					
COMM. TO MAIN #	⌀ BRANCH CONN.	TOTAL ⌀ BRANCH	BRANCH SIZE	TOTAL ⌀ MAIN	MAIN SIZE

FORM F

JOB _____ DATE: _____ BY _____

DRAINAGE, SIZING					
WING	STACK OR FIXTURES	FIXTURE UNITS BR.	TOTAL F.U. MAIN	SIZE	REMARKS

FORM G   Sheet 1

```
JOB_____          SHEET NO. 1 OF ____ SHEETS
SYSTEM_____          BY_____ DATE_____
```

DOMESTIC WATER SYSTEM

AVAILABLE: _____ * PSI AT ELEV. _____ IN _____ STREET

Total St. Pr. FU _____
St. Pr. GPM   _____
Pump GPM   _____
HVAC GPM   _____

Total GPM   _____
Size   _____
Frict. Loss/100' _____
Allowance for future drop in St. Pr. _____ PSI

① * Adjusted Available St. Pr. _____ PSI x 2.31 = _____ ft.

② SERVICE LOSSES

No.	Size	Item

Gate Valves   @ _____ = _____ ft.
Check Valves   @ _____ = _____ ft.
90° Els (Misc Ftgs) @ _____ = _____ ft.
                     Run   _____ ft.

Total Footage = _____ ft.

Total Footage x Loss/100' = Fr Loss in Ft.

_____ x _____ = _____ ft.
_____ " Strainer Loss @ _____ PSI x 2.31 = _____ ft.
_____ " Meter   Loss @ _____ PSI x 2.31 = _____ ft.

Total Friction Loss For Service [_____ ft.] ②

SERVICES LOSSES (TYPICAL)

FORM G   Sheet 2

```
JOB_____          SHEET NO. 2  OF_____ SHEETS
SYSTEM_____          BY_____ DATE_____
```

DIRECT FEED STREET PRESSURE SYSTEM

③ LIFT

    Elevation of top most fixture _____ ft.
    Minus Elevation of St. Main - _____ ft.

                         LIFT [_____] ③

④ HEAD

    Pressure Reqd. at top most fixture _____ PSI x 2.31 _____ ft.
    Service Losses ②   _____ ft.
    Lift ③   _____ ft.

                      Reqd. Head (Less Fr.) [_____] ④

    Adjusted Avail. St. Press. ① _____ ft.
    Minus Reqd. Head ④   - _____ ft.

                Pressure Remaining [_____] ft.
                    For Friction

CONCLUSIONS:

FORM G   Sheet 3

```
JOB_____          SHEET NO. 3  OF_____SHEETS
SYSTEM_____          BY_____DATE_____
```

_____ PUMP CALCULATION

③ LIFT

    Elevation one (1) ft. above house Tk.⎱
    Elevation of top most fixture   ⎰ _____ ft.
    Minus Elev. of Pump Room floor _____ ft.

                          LIFT [_____ ft.] ③

④ Pump Suction Losses (After Service)*

No.	Size	Items			
		Gate Valves	@ _____	= _____	ft.
		Check Valves	@ _____	= _____	ft.
		90° Els (Misc Ftgs)	@ _____	= _____	ft.
		Run		_____	ft.

                    Total Footage = _____ ft.

Total Footage x Fr. Loss/100' = Fr. Loss in Suction

_____ x _____ = [_____ ft.] ④

⑤ Pressure Available At Pump

    Adjusted St. Pr. ①   _____ ft.
    Gain to Pump Room Fl. Elev. _____ ft.
                              _____ ft.

Minus ⎰ Service Losses ② _____ ft.
       ⎱ Suction Losses ④ _____ ft.

                   _____ ft. ← _____ ft.

             Available Pressure At Pump [_____ ft.] ⑤

* NOTE: For complicated suction piping, additional calculations will be required. Use main sizing sheets.

FORM G Sheet 4

```
JOB_____          SHEET NO. 4  OF ____ SHEETS
SYSTEM_____          BY_____ DATE_____
```

⑥ Pump Discharge (System) Losses

No.	Size	Item			
		Gate Valves	@ _____	= _____	ft.
		Check Valves	@ _____	= _____	ft.
		90° Els (Misc Ftgs)	@ _____	= _____	

Total Footage = _____ ft.

Total Footage × Fr. Loss/100' = Total Fr. Loss in Discharge

_____ × _____ = _____ ft. ⑥

HEAD

Pressure Reqd. at top of System ____ PSI × 2.31 = _____ ft.
Lift ③ _____
Pump Discharge Losses ⑥ _____
Total _____
Minus Adjusted Avail. Pressure At Pump ⑤ _____

REQUIRED HEAD _____

CONCLUSIONS:

FORM H

				COLD WATER								HOT WATER						
FLOOR	PRESSURE		F.U. DRAIN	F.U. BR.	G.P.M. BR.	SIZE BR.	F.U. RISER	G.P.M. RISER	SIZE RISER	FR. FT/100	FR. LOSS	F.U. BR.	G.P.M. BR.	SIZE BR.	F.U. RISER	G.P.M. RISER	SIZE RISER	FR. FT/100	FR. LOSS

RISER SIZING – STACK N°

HEIGHT OF SOIL   SOIL STACK   F.U.
HEIGHT OF VENT   VENT STACK   F.U.

FORM I

				COLD WATER								HOT WATER						
NO.	RUN		F.V. DRAIN	F.V. BR.	G.P.M. BR.	SIZE BR.	F.V. MAIN	G.P.M. MAIN	SIZE MAIN	FR. FT/100	FR. LOSS	F.V. BR.	G.P.M. BR.	SIZE BR.	F.V. MAIN	G.P.M. MAIN	SIZE MAIN	FR. FT/100	FR. LOSS

MAIN SIZING – SHEET Nº

FORM J

HOT WATER RECIRCULATION SYSTEM SIZING FORM

CAL. GPM CIR PUMP CAPACITY												CAL. FT HEAD OF PUMP					
SUPPLY						RETURN						TOTALS		RETURN			
SECT	SUB SECT	DIA	MEAS L	HT. LOSS RATE	HEAT LOSS	SECT	SUB SECT	DIA	MEAS L	HT. LOSS RATE	HEAT LOSS	TOT. HEAT LOSS	Q = ** H.L./5000	EQUIV LGTH. 1½×L	FR/100 ft	FR LOSS	TOT GOV FR*
TOTALS:–																	

*Total friction loss (ft) from longest or governing run = pump head.
**Total of column = gpm capacity of pump.

FORM K

JOB _____ DATE _____ BY _____

GAS RISER SIZING										
FLOOR		OUTLETS BR	C.F.H. BR	SIZE BR	OUTLETS RISER	DEMAND FACTOR	C.F.H. RISER	SIZE RISER	PRESS LOSS in/100	PRESS. LOSS

FORM L

JOB _____ DATE _____ BY _____

			GAS MAIN SIZING							
NO.	EQUIV. RUN	OUTLETS BR.	C.F.H. BR.	SIZE BR.	EQUIP/ OUTLETS MAIN	DEMAND FACTOR	C.F.H. MAIN	SIZE MAIN	PRESS. LOSS in/100	PRESS. LOSS

FORM M

JOB _____ DATE _____ BY _____

LABORATORY GAS SIZING SHEET NO. _____
LOCATION _____
TOTAL LOSS IN MAIN BEYOND BRANCH CONN _____
_____ CFH/BUNSEN BURNER

NO.	ACTUAL RUN	FITTINGS	EQUIV. RUN	CFH BRANCH	SIZE BRANCH	TOTAL CFH	FACTOR	ADJ. CFH	SIZE	LOSS	TOTAL LOSS	

FORM N

JOB _____ DATE _____ BY _____

NO.	RUN	FITTING	EQUIV. RUN	CONN. CFM BRANCH	SIZE BRANCH	TOTAL MAIN	FACTOR	ADJ. CFM	SIZE	FR. psi/100 ft	FR. LOSS	NO.	RUN	FITTING	EQUIV. RUN	CONN. CFM BRANCH	SIZE BRANCH	TOTAL MAIN	FACTOR	ADJ. CFM	SIZE	FR. psi/100 ft	FR. LOSS

AIR SIZING – SHEET NO.

FORM O

JOB _____ DATE _____ BY _____

				1 FACTOR ___%				2 FACTOR ___%				3 FACTOR ___%				ADJ BRANCH CFM				TOTAL ADJ. CFM (TOTAL FROM 1,2,3)			
CONN NO.	RUN	FITTINGS	EQUIV RUN	BRANCH CFM	BRANCH SIZE	TOTAL CFM	ADJ. CFM	BRANCH CFM	BRANCH SIZE	TOTAL CFM	ADJ. CFM	BRANCH CFM	BRANCH SIZE	TOTAL CFM	ADJ. CFM	NO. @ ___%	NO. @ ___%	NO. @ ___%	TOTAL ADJ. CFM		SIZE	FR. in Hg/100 ft	FRICTION LOSS

VACUUM SIZING SHEET NO.

5-21

FORM P

JOB _____ DATE _____ BY _____

						OXYGEN SIZING SHEET NO.							
NO.	RUN	TOTAL RUN	OUTLETS	TOTAL OUTLETS	ADJ. LTR/MIN	SIZE	NO.	RUN	TOTAL RUN	OUTLETS	TOTAL OUTLETS	ADJ. LTR/MIN	SIZE

5-22

FORM Q

JOB _____ DATE _____ BY _____

NITROUS OXIDE SIZING SHEET NO.														
NO.	RUN	TOTAL RUN	OUTLETS	TOTAL OUTLETS	ADJ. LTR/MIN	SIZE	NO.	RUN	TOTAL RUN	OUTLETS	TOTAL OUTLETS	ADJ. LTR/MIN	SIZE

APPENDIX

Units of Measure and SI Conversion factors

The following conversion factors are appropriate for the units of measure used in this Plumbing Design Guide:

Length

1 inch (in) = 0.0254 meter (m)
1 foot (ft) = 0.3048 meter (m)

Area

1 square foot (ft^2) = 0.0929 $meter^2$ (m^2)

Volume

1 U.S. liquid gallon (gal) = 3.785 liters (L)
1 cubic foot (ft^3) = 0.0283 $meter^3$ (m^3)

Velocity

1 foot per second (fps) = 0.305 meter/second (m/s)
1 foot per minute (fpm) = 0.00508 meter/second (m/s)

Flow rate

1 U.S. liquid gallon per minute (gpm) = 0.0631 liter/second (L/s)
1 cubic foot per minute (cfm) = 0.0004719 $meter^3$/second (m^3/s)

Pressure

1 pound force per square inch (psi) = 6.895 kilopascals (kPa)
1 inch of Mercury (Hg) = 3.377 kilopascals (kPa)
1 inch of water = 248.84 pascals (Pa)

Miscellaneous

1 Btu = 1.055 kilojoules (kJ)
1 electric horsepower (hp) = 745.700 watts (W)
1 pound (lb) = 0.454 kilogram (kg)

Temperature

C = (F-32)/1.8
F = (1.8 x C) + 32

INDEX

Access doors, 4-4
Acid-neutralizing sumps, 4-25 to 4-27
Air chambers, 4-33
Air compressors, 4-162, 4-163
Air dryers, 4-163, 4-164
Air relief valves, 4-33, 4-34
Alarms:
  for carbon-dioxide fire-extinguishing systems, 4-152
  for compressed-air systems, 4-159
  for dry-chemical fire-extinguishing systems, 4-154
  for ejector pits, 4-23
  for fire standpipe systems, 4-141
  for foam fire-extinguishing systems, 4-154
  for gravity tanks, 4-54 to 4-56
  for Halon fire-extinguishing systems, 4-154
  for medical gas systems, 4-183
  for sprinkler systems, 4-148
  for sump pits, 4-16
  for vacuum air systems, 4-170
Arm baths, 4-209, 4-210

Backflow prevention:
  for bathtubs, 4-206
  for can washers, 4-209
  for decorative fountains, 4-124
  for domestic water supply, 4-31, 4-35
  for fire standpipe systems, 4-139
  for irrigation systems, 4-104
  for laboratory trim, 4-210
  for mop sinks, 4-204, 4-205
  for showers, 4-207
  for slop sinks, 4-204
  for sprinkler systems, 4-147
  for swimming pools, 4-111
  for urinals, 4-200
Backwater valves, 4-14, 4-19
Balancing (surge) tanks, 4-110
Bathtubs, 4-206, 4-207
Bedpan washers, 4-208
Boosted-pressure water systems:
  design principles of, 4-48 to 4-51
  equipment for, 4-51 to 4-64

Calculations, 2-3
Can washers, 4-36, 4-208, 4-209
Carbon-dioxide fire-extinguishing systems,
   design principles of, 4-152 to 4-154
Catch basins, 3-14
Check list for completion of project, 1-4
Chilled-drinking-water systems:
  design principles of, 4-96, 4-97
  equipment for, 4-97 to 4-99
  recirculation, 4-98
Cleanouts, 4-10
Clinical (flushing-rim) sinks, 4-204
Completion of project, check list for, 1-4
Compressed-air outlets, 4-157, 4-159
Compressed-air systems, 4-160
  design principles of:
    building systems, 4-160
    dental systems, 4-159, 4-160
    hospital systems, 4-157 to 4-159
    laboratory systems, 4-156, 4-157

Compressed-air systems (Cont.):
  equipment for, 4-162 to 4-166
  pipe sizing for, 4-157 to 4-162
  piping materials for, 4-162
Conformity with requirements, 2-1
  buildings:
    carbon-dioxide fire-extinguishing systems, 4-152
    dental compressed-air systems, 4-159
    domestic water systems, 4-30
    drainage, 4-9
    dry-chemical fire-extinguishing systems, 4-155
    fire standpipe systems, 4-137
    foam fire-extinguishing systems, 4-155
    Halon fire-extinguishing systems, 4-154
    hospital compressed-air systems, 4-157
    medical gas systems, 4-181 1
    sprinkler systems, 4-146
    swimming pools, 4-109
  site, 3-1
    drainage, 3-6
    fire protection, 3-24
    gas, 3-26
    water supply, 3-19
Contraction and expansion of
  piping, 4-3, 4-4, 4-8, 4-10, 4-34
Controlled flow for roof drainage, 4-11 to 4-13
Cross connections, 3-21, 4-31

Decorative pools and fountains:
  design principles of, 4-117 to 4-126
  equipment for, 4-128 to 4-132
  fittings for, 4-122 to 4-124
  piping materials for, 4-128
Demineralized (deionized) water systems:
  design principles of, 4-100, 4-101
  equipment for, 4-101, 4-102
  piping materials for, 4-101
Demineralizers (deionizers), 4-101, 4-102
Design principles, 2-1 to 2-5
  building, 4-1 to 4-8
  of carbon-dioxide fire-extinguishing systems, 4-152
    to 4-154
  of central soap systems, 4-191, 4-192
  of compressed-air systems:
    building systems, 4-160
    dental systems, 4-159, 4-160
    hospital systems, 4-157 to 4-159
    laboratory systems, 4-156, 4-157
  of decorative pools and fountains, 4-117 to 4-126
  drainage:
    building, 4-9, 4-10
      garage, 4-27, 4-28
      infectious waste, 4-29, 4-30
      laboratory waste, 4-24, 4-25
      radioactive waste, 4-29
      sanitary, 4-18 to 4-20
      storm water, 4-11 to 4-15
    site, 3-6
      sanitary, 3-15, 3-16
      storm water, 3-13 to 3-15
  of dry-chemical fire-extinguishing systems, 4-155
  of fire extinguishers, 4-151, 4-152
  of fire standpipe systems, 4-137 to 4-141
  of foam fire-extinguishing systems, 4-155

Design principles (Cont.):
 of gas systems:
  buildings, 4-132 to 4-135
  site, 3-26
 of gasoline systems, 4-193
 of Halon fire-extinguishing systems, 4-154, 4-155
 insulation, 4-193 to 4-195
 of irrigation systems, 4-104 to 4-108
 of medical gas systems, 4-181 to 4-183
 of nitrogen systems, 4-190
 of nitrous oxide systems, 4-188
 of oral vacuum systems, 4-174
 of oxygen systems, 4-183, 4-184
 site, 3-1, 3-2
 of sprinkler systems, 4-145 to 4-148
 of swimming pools, 4-109 to 4-112
 of vacuum air systems, 4-167 to 4-170
 of vacuum cleaning systems, 4-177, 4-178
 of vibration isolation, 4-197
 of water supply systems:
  building, 4-30 to 4-36
   boosted-pressure systems, 4-48 to 4-51
   chilled-drinking-water systems, 4-96, 4-97
   demineralized (deionized) water systems, 4-100, 4-101
   distilled-water systems, 4-99
   hot-water systems, 4-65 to 4-67
   nonpotable-water systems, 4-104
   reverse osmosis (RO) water systems, 4-102
   street-pressure systems, 4-48
   water treatment, 4-102
  site, 3-19 to 3-22
   domestic, 3-22, 3-23
   fire protection, 3-23 to 3-25
Design procedures, 1-1 to 1-4
Distilled-water systems:
 design principles of, 4-99
 equipment for, 4-99, 4-100
 piping materials for, 4-99
Drainage systems, building: design principles
   of (see Design principles, drainage, building)
   garage, 4-27, 4-28
   highly infectious waste, 4-29, 4-30
   laboratory waste, 4-24
   radioactive waste, 4-29
   sanitary, 4-18
   storm water, 4-11
Drinking fountains (water coolers):
 types of, 4-205, 4-206
 where required, 4-96
Dry-chemical fire-extinguishing systems, design
   principles of, 4-155

Electric motors and starters, 2-4
Emergency-use fixtures, 4-208
Energy and water conservation possibilities, 2-8,
   2-9, 4-1
 air compressors, 4-163
 fixtures, 4-199
 lavatories, 4-202
 sewage ejectors, 4-19
 showers, 4-207
 vacuum air pumps, 4-173
Expansion and contraction of piping, 4-3, 4-4, 4-8,
   4-10, 4-34

Facilities for the handicapped, 4-198
  drinking fountains, 4-96, 4-205, 4-206
  lavatories, 4-201
  water closets, 4-200
Filters:
  for chilled-drinking-water systems, 4-98, 4-99
  for compressed-air systems, 4-164
  for decorative fountains, 4-128, 4-129
  for swimming pools, 4-113, 4-114
  for vacuum air systems, 4-173, 4-174
  for water treatment, 4-103
Fire department connections (see Siamese fire
  department connections)
Fire extinguishers, 4-151, 4-152
  (See also specific fire-extinguishing system)
Fire hydrants, 3-21, 3-22, 3-24
Fire-line systems, site:
  design principles of, 3-23 to 3-25
  pipe sizing for, 3-21
  piping materials for, 3-25
Fire-protection systems, building:
  carbon-dioxide fire-extinguishing systems, 4-152
    to 4-154
  dry-chemical fire-extinguishing systems, 4-155
  fire extinguishers, 4-151, 4-152
  fire standpipe systems, 4-137 to 4-145
  foam fire-extinguishing systems, 4-155
  Halon fire-extinguishing systems, 4-154, 4-155
  sprinkler systems, 4-145 to 4-151
Fire standpipe systems:
  design principles of, 4-137 to 4-141
  equipment for, 4-142 to 4-145
  pipe sizing for, 4-139, 4-140
  piping materials for, 4-141, 4-142
Fixture supports:
  drinking fountains (water coolers), 4-205
  flushing-rim clinical sinks, 4-204
  laundry trays, 4-205
  lavatories, 4-202, 4-203
  sinks, 4-204
  urinals, 4-201
  water closets, 4-200
Flexible connections, 4-196, 4-197
Float valves, 4-34, 4-35, 4-54
Foam fire-extinguishing systems, design
  principles of, 4-155
Fountain nozzles, 4-119 to 4-122
Fountains (see Decorative pools and fountains;
  Drinking fountains)

Garage drainage systems:
  design principles of, 4-27, 4-28
  equipment for, 4-28
  piping materials for, 4-28
Gas systems:
  building:
    design principles of, 4-132 to 4-135
    pipe sizing for, 4-135
    piping materials for, 4-137
  site:
    design principles of, 3-26
    pipe sizing for, 3-27
    piping materials for, 3-27
Gasoline systems, design principles of, 4-193

Gravity (house) tanks:
  domestic, 4-54 to 4-56
  fire standpipe, 4-138, 4-144
  sprinkler, 4-147, 4-151
Grease traps, 4-23, 4-24

Halon fire-extinguishing systems, design
  principles of, 4-154, 4-155
Handicapped, the (see Facilities for the
  handicapped)
High-rise building drainage, 4-18
Highly infectious waste-water drainage systems:
  design principles of, 4-29, 4-30
  piping materials for, 4-30
Hose bibbs, 4-35
Hot-water branches, dead-leg, maximum
  allowable lengths of, 4-76
Hot-water recirculation, where required, 4-66
Hot-water systems:
  design principles of, 4-65 to 4-67
  equipment for, 4-67 to 4-73
House tanks [see Gravity (house) tanks]
House traps, 4-14, 4-20
Hubbard tanks (full-body bath), 4-209, 4-210
Hydrant flow tests:
  domestic, 4-30
  fire standpipe, 4-137, 4-138
  site, 3-9, 3-19
  sprinkler, 4-146
Hydropneumatic pressure tanks:
  domestic, 4-58
  fire standpipe, 4-138, 4-144, 4-145
  sprinkler, 4-147
Hydrotherapeutic equipment, 4-209, 4-210
Hydrotherapeutic pools, 4-109, 4-209
Hypochlorinators:
  for fountains, 4-125, 4-131, 4-132
  for swimming pools, 4-116, 4-117

Information required at the start of a project, 1-1
Insulation:
  design principles of, 4-193 to 4-195
  for drainage piping, 4-194
  for fountains, 4-126, 4-194, 4-195
  for swimming pools, 4-112, 4-194
  for water piping, 4-33, 4-193 to 4-195
Irrigation systems:
  design principles of, 4-104 to 4-108
  pipe sizing for, 4-108
  piping materials for, 4-108

Laboratory trim, 4-210, 4-211
Laboratory waste-water drainage systems:
  design principles of, 4-24, 4-25
  equipment for, 4-25 to 4-27
  piping materials for, 4-24, 4-25
Laundry trays, 4-205
Lavatories, 4-201 to 4-203
Leg baths, 4-209, 4-210
Low water cutoffs, 4-51

Manholes, 3-6, 3-14
Medical gas outlets, 4-183

Medical gas systems:
  alarm systems for, 4-183
  design principles of, 4-181 to 4-183
  piping materials for, 4-183
Mop sinks, 4-204, 4-205

Nitrogen systems:
  design principles of, 4-190
  outlets for, 4-191
  pipe sizing for, 4-190
Nitrous oxide systems:
  design principles of, 4-188
  pipe sizing for, 4-189
Nonpotable-water systems:
  design principles of, 4-104
  equipment for, 4-104
  piping materials for, 4-104

Oil separators, 4-14, 4-28
Oral vacuum systems:
  design principles of, 4-174
  equipment for, 4-177
  pipe sizing for, 4-174, 4-175
  piping materials for, 4-175
Oxygen systems:
  design principles of, 4-183, 4-184
  pipe sizing for, 4-187

Pipe sizing:
  for compressed-air systems, 4-157 to 4-162
  for drainage systems:
    building, storm-water drainage, 4-13
    site, storm-water drainage, 3-13, 3-14
  for fire-line systems, 3-21
  for fire standpipe systems, 4-139, 4-140
  for gas systems:
    building, 4-135
    site, 3-27
  for irrigation systems 4-108
  for nitrogen systems, 4-190
  for nitrous oxide systems, 4-189
  for oral vacuum systems, 4-174, 4-175
  for oxygen systems, 4-187
  for soap systems, 4-192
  for sprinkler systems, 4-149
  for vacuum air systems, 4-170 to 4-172
  for vacuum cleaning systems, 4-178, 4-179
  for water systems, 4-36 to 4-37, 4-46 to 4-47
Piping:
  expansion and contraction of, 4-3, 4-4, 4-8, 4-10, 4-34
  subject to freezing, 3-21, 4-1, 4-2, 4-10, 4-32
Piping materials:
  for compressed-air systems, 4-162
  for drainage systems:
    building:
      garage, 4-28
      infectious waste, 4-30
      laboratory waste, 4-25
      radioactive waste, 4-29
      sanitary, 4-20, 4-21
      storm water, 4-15

Piping materials, for drainage systems (Cont.):
  for site:
    sanitary, 3-16
    storm water, 3-15
  for fire-line systems, 3-25
  for fire standpipe systems, 4-141, 4-142
  for fountain systems, 4-128
  for gas systems:
   building, 4-137
   site, 3-27
  for irrigation systems, 4-108
  for medical gas systems, 4-183
  for oral vacuum systems, 4-175
  for soap systems, 4-192
  for sprinkler systems, 4-148, 4-149
  for swimming pool systems, 4-112, 4-113
  for vacuum air systems, 4-172
  for vacuum cleaning systems, 4-179
  for water supply systems:
   building:
    demineralized (deionized water), 4-101
    distilled water, 4-99
    domestic water, 4-37, 4-38
    reverse osmosis (RO) water, 4-102
   site, 3-23
Plumbing fixtures, 4-198, 4-199
 bathtubs, 4-206, 4-207
 bedpan washers, 4-208
 can washers, 4-208
 drinking fountains (water coolers), 4-205, 4-206
 emergency-use fixtures, 4-208
 flushing-rim clinical sinks, 4-204
 hydrotherapeutic equipment, 4-209 to 4-212
 laundry trays, 4-205
 lavatories, 4-201 to 4-203
 mop sinks, 4-204, 4-205
 showers, 4-207
 sinks, 4-203, 4-204
 slop sinks, 4-204
  space required behind, 4-3, 4-5 to 4-7
  supports for (see Fixture supports)
 urinals, 4-200, 4-201
 water closets, 4-199, 4-200
Pneumatic sewage ejectors, 4-9, 4-19, 4-21, 4-22
Pools (see Decorative pools and fountains;
  Swimming pools)
Preliminary systems loads, 2-5 to 2-8
Pressure gauges, 4-4
 for fountains, 4-125
 for swimming pools, 4-111
Pressure-reducing valves:
 for compressed air, 4-156
 for domestic water, 4-32, 4-33
 for fire standpipe, 4-139
 for sprinklers, 4-148
Pumps:
 chilled-drinking-water recirculating, 4-98
 domestic water booster, 4-48, 4-50 to 4-53
 fire, sprinkler, 4-147, 4-149 to 4-151
 fire standpipe, 3-24, 4-142 to 4-144
 fountain, 4-129, 4-130
 hot-water recirculating, 4-71 to 4-73
 house, 4-52 to 4-54
 hydropneumatic tank:
  domestic, 4-57
  fire standpipe, 4-144
  sprinkler, 4-151

Pumps (Cont.):
 jockey:
  fire standpipe, 4-138, 4-143
  sprinkler, 4-147, 4-150
 sewage ejector, 3-16 to 3-18, 4-22, 4-23
 sump, 4-15 to 4-17
 swimming pool, 4-114, 4-115
 tank fill:
  domestic, 4-52 to 4-54
  fire standpipe, 4-145
  sprinkler, 4-151
 well, 3-20

Radioactive waste-water systems:
 design principles of, 4-29
 piping materials for, 4-29
Rate-of-flow indicators:
 for fountains, 4-119, 4-125
 for swimming pools, 4-111
Relief valves:
 for hot-water heaters and tanks, 4-68
 for hydropneumatic tanks, 4-57
 hydrostatic, for swimming pools, 4-112
Relief vents, drainage, 4-19
Reverse osmosis (RO) water systems:
 design principles of, 4-102
 equipment for, 4-102
 piping materials for, 4-102

Sanitary drainage systems, building:
 design principles of, 4-18 to 4-20
 equipment for, 4-21 to 4-24
 piping materials for, 4-20, 4-21
Sanitary sewer systems:
 design principles of, 3-15, 3-16
 piping materials for, 3-16
Service sinks, 4-204
Sewer pipes:
 available sizes of, 3-9
 capacities of, 3-6 to 3-8
 soil loading, 3-10 to 3-12
Shock absorbers, 4-33
Showers, 4-207
Siamese fire department connections:
 fire standpipe, 4-138
 site, 3-22
 sprinkler, 4-145
Sill cocks, 4-35
Sink trim, 4-211
Sinks, 4-203, 4-204
Slop (service) sinks, 4-204
Soap systems:
 design principles of, 4-191, 4-192
 individual dispensers for, 4-193
 outlets for, 4-193
 piping for, 4-192
 soap tanks for, 4-192, 4-193
Space required behind fixtures, 4-3, 4-5 to 4-7
Specifications, 2-2
Spray heads for irrigation systems, 4-104 to 4-10
Sprinkler systems:
 design principles of, 4-145 to 4-148
 equipment for, 4-149 to 4-151

Sprinkler systems (Cont.):
  pipe sizing for, 4-149
  piping materials for, 4-148, 4-149
Steam-water mixing valves, 4-67
Stills, 4-99, 4-100
Storm-water drainage systems, building:
  design principles of, 4-11 to 4-15
  equipment for, 4-15 to 4-17
  piping materials for, 4-15
Storm-water sewer systems:
  design principles of, 3-13 to 3-15
  pipe sizing for, 3-13, 3-14
  piping materials for, 3-15
Street-pressure water supply systems, design principles of, 4-48
Structural information, 2-4
Suction tanks, 4-51
Sump pumps, 4-15 to 4-17
Surge (balancing) tanks, 4-110
Swimming pools:
  design principles of, 4-109 to 4-112
  equipment for, 4-113 to 4-117
  fittings for, 4-110 to 4-112
  piping materials for, 4-112, 4-113

Tamper switches:
  fire standpipe, 4-139
  sprinkler, 4-148
Temperatures of water:
  for can washers, 4-209
  chilled drinking water, 4-96
  hot water, 2-6, 4-65, 4-67
  hydrotherapeutic equipment, 4-210
  in hydrotherapeutic pools, 4-109
  in swimming pools, 4-111
Thermometers, 4-4
Toilet accessories, 4-211, 4-212

Urinals, 4-200, 4-201
Utility information, 3-3 to 3-6, 3-9, 4-1, 4-9

Vacuum air inlets, 4-170
Vacuum air pumps, 4-173
Vacuum air systems:
  design principles of, 4-167 to 4-170
  equipment for, 4-173, 4-174
  pipe sizing for, 4-170, 4-171
  piping materials for, 4-172
Vacuum cleaning:
  fountain, 4-126
  pool: outlets for, 4-111
    pumps for, 4-114
    tools for, 4-117
Vacuum cleaning systems:
  design principles of, 4-177, 4-178
  equipment for, 4-181
  pipe sizing for, 4-178, 4-179
  piping materials for, 4-179
Vacuum producers:
  oral vacuum, 4-177
  vacuum cleaning, 4-181

Velocity, piping:
  compressed air, 4-161
  fountains, 4-126
  sanitary drainage, 3-16
  storm-water drainage, 3-14
  swimming pools, 4-112
  vacuum air, 4-171
Vibration isolation, design principles of, 4-195 to 4-198

Wash sinks, 4-203
Water closets, 4-199, 4-200
Water conservation (see Energy and water conservation possibilities)
Water coolers (see Drinking fountains)
Water-hammer arresters, 4-33
Water heaters:
  domestic, 4-65
    sizing data for, 4-77 to 4-95
    types of, 4-67 to 4-73
  for fountains, 4-130, 4-131
  for laundry waste heat reclaimers, 4-70
  preheaters, 4-70, 4-71
  for swimming pools, 4-115, 4-116
Water meters, 3-21, 4-31
Water pressures, 4-32
  for can washers, 4-209
  for water closets, 4-199, 4-200
Water softeners, 4-103
Water storage tanks, site:
  domestic, 3-22
  fire, 3-24
Water supply systems:
  building, 4-30 to 4-36
    boosted pressure, 4-48 to 4-51
    chilled drinking water, 4-96, 4-97
    demineralized (deionized) water, 4-100, 4-101
    distilled water, 4-99
    hot water, 4-65 to 4-67
    nonpotable water, 4-104
    pipe sizing for, 4-36 to 4-37, 4-46 to 4-47
    piping materials for, 4-37
    reverse osmosis (RO) water, 4-102
    water treatment, 4-102
  site:
    design principles of, 3-19 to 3-22
      domestic, 3-22, 3-23
      fire protection, 3-23 to 3-25
    domestic booster pumping stations, 3-23
    fire pumping stations, 3-24
    on-site storage, 3-22, 3-24
    piping materials for:
      domestic, 3-23
      fire protection, 3-25
Water treatment:
  design principles of, 4-102
  equipment for, 4-103
  piping materials for, 4-102
Waterfalls:
  design data for, 4-121, 4-122
  splash of, 4-118
Wells and well pumps, 3-20
Wind controls for fountains, 4-125